Preface

The development of cell culture and its application in the scientific field is based on first class techniques; no other standard is acceptable. A good theoretical knowledge will not on its own produce good cultures. It is the skill acquired by practice and, I believe, a sympathy for the subject that produces satisfactory results.

The purpose of this book is to bring together in one volume some of the developments in method which have taken place in recent years and are now used in cell culture. It is hoped that the information given will be of practical use to laboratory workers who are using cell cultures in their scientific work.

I would like to express my gratitude to all the contributors who have been most patient and helpful in the publication of this book.

Acknowledgements are due to the various organisations who have kindly given their permission to reproduce the various illustrations and photographs in this book.

G.D.W.

Animal Tissue Culture

Advances in Technique

Animal Tissue Culture
Advances in Technique

Edited by
GERALD D. WASLEY

LONDON
BUTTERWORTHS

A/599.08

THE BUTTERWORTH GROUP

ENGLAND
Butterworth & Co. (Publishers) Ltd.
London: 88 Kingsway, WC2B 6AB

AUSTRALIA
Butterworth & Co. (Australia) Ltd.
Sydney: 586 Pacific Highway Chatswood, NSW 2067
Melbourne: 343 Little Collins Street, 3000
Brisbane: 240 Queen Street, 4000

CANADA
Butterworth & Co. (Canada) Ltd.
Toronto: 14 Curity Avenue, 374

NEW ZEALAND
Butterworth & Co. (New Zealand) Ltd.
Wellington: 26–28 Waring Taylor Street, 1

SOUTH AFRICA
Butterworth & Co. (South Africa) (Pty) Ltd.
Durban: 152–154 Gale Street

First published 1972

© Butterworth & Co. (Publishers) Ltd.

ISBN 0 408 70340 7

Printed in England by Page Bros (Norwich) Ltd., Norwich and London

Contents

ONE

Mammalian cell culture media

A. M. WHITAKER

Department of Virology, Wellcome Research Laboratories, Beckenham, Kent

GENERAL CONSIDERATIONS

The design of media for cell culture creates two conflicting interests which are difficult to reconcile. The aim must be to provide an environment for the cell which is as close as possible to that experienced *in vivo*, yet in order to obtain some measure of control and standardisation over the conditions it is also necessary to define each of the constituents. Whereas natural media, body fluids, tissue extracts etc., fulfil the conditions of the former they fall far short of fulfilling the latter. Likewise, media made up of purified and characterised components bear little resemblance to natural conditions.

The problem is most easily solved for short-term survival, since the only factors of importance are osmotic pressure, hydrogen ion concentration, other inorganic ions, source of carbohydrate and gases. Combinations of this type are known as balanced salt solutions. The osmotic pressure of a normal mammalian cell is in the region of 7·6 atmospheres at 38°C corresponding to a depression of freezing point of about 0·63°C. Although cells are not greatly affected by changes in the osmotic pressure as large as ±10 per cent, or even greater providing the change takes place slowly, it is important to keep close to the normal. Sodium chloride makes the greatest contribution to the osmotic pressure in animal cells and in most cell culture media, with other inorganic ions and glucose also playing their part. High molecular weight substances make relatively little contribution. It is therefore easier, from the point of maintaining a stable osmotic pressure, to vary the concentration of a protein in the medium rather than that of a small molecule such as glucose.

The hydrogen ion concentration of medium must be kept close to neutrality, even though most tissues will tolerate quite wide divergencies from pH 7·0. Optimal growth normally occurs between pH 7·2 and 7·4 and for certain cells,

1

for example human diploid cell lines, it is important not to exceed these limits. Survival of most other tissues is not likely to persist much beyond pH 6·8 and 7·6. For many years the pH in culture media has been controlled by a buffer system modelled on the naturally occurring CO_2/bicarbonate system present in blood plasma. The bicarbonate is added to the medium as part of the balanced salt solution, which normally includes a weak phosphate buffer. Although this system works quite satisfactorily and is used in the majority of culture media, it has a number of disadvantages. CO_2 is readily lost from the medium, both during storage and use, with a consequent rise in pH. In a closed culture vessel, cell respiration usually provides sufficient CO_2 and acid metabolites to maintain the pH at a satisfactory level. During the initiation of a culture, however, the pH can rise considerably. The problem may be overcome by bubbling CO_2 through the medium prior to use, or by incubating the cultures in open vessels in an incubator with a CO_2 enriched atmosphere.

In addition to these difficulties, the pKa of sodium bicarbonate is 6·1, and this results in suboptimal buffering over the physiological range. These disadvantages have stimulated a number of attempts to find a non-toxic, non-volatile buffer system. Glycylglycine[1] and *tris*-(hydroxymethyl) aminomethane[2] have been used for some time, although Martin[3] has more recently stressed the importance of adjusting the pH of *tris*- buffers with HCl rather than with the more toxic citric or malic acids. He found good proliferation of cells when this buffer was used in a simple chemically defined medium. Nevertheless, several workers have reported that both glycylglycine and *tris*- buffers are toxic over long periods and Shipman[4] claims that 4-(2-Hydroxyethyl)-1-piperazine ethane sulphonic acid (HEPES) has considerable advantages over them. HEPES is non-toxic in all systems tested so far, exhibits no metal binding and is readily soluble at 0°C. A wide variety of cells grew well in its presence without any effect on subsequent virus titrations or haemagglutination. In 1963, Leibowitz[5] devised a medium for use in free gas exchange with air. Free-base amino acids, especially L-arginine, L-cysteine and L-histidine, were substituted for sodium bicarbonate, while acid production was reduced by substituting more oxidised carbohydrate sources for glucose.

Other inorganic ions required for survival include sodium, potassium, calcium, magnesium, iron, carbonate, phosphate and sulphate. Their function, although not fully understood, includes maintenance of osmotic pressure, contribution to enzyme and metabolic activity and formation of the cell-to-glass bond. Glucose is the most common carbohydrate source and is a constituent of most balanced salt solutions. Some of the more complex media contain other sugars or simpler compounds including lactic, pyruvic and acetic acids in addition to, or instead of glucose. Oxygen and carbon dioxide are almost certainly essential for survival. For most cultures, dissolved oxygen in the medium is sufficient, but it is important to ensure that the ratio of air-space to medium in closed vessels is large enough to supply an adequate concentration. Carbon dioxide is produced by metabolism, so that its exclusion from medium and thus its requirement is difficult to establish. There is, however, strong evidence to suggest that some cells do have an absolute requirement.

Media become correspondingly more complex for long-term survival and growth. Balanced salt solutions will maintain cells in a healthy condition for only a relatively short period, but this can be considerably extended by the addition of a cell protective agent such as serum, gelatin, albumin or methyl cellulose.[6] However, in order to develop the full potential and function of the cell, a much wider range of medium components is required so that new cellular material can be produced and metabolic activity increased. Amino acids are important among these. In addition to the ten 'essential' amino acids required by mammals, (arginine, histidine, isoleucine, leucine, lysine, methionine, phenylalanine, threonine, tryptophane and valine), mammalian cells in culture also require cystine and tyrosine[7]. Most cells, in particular transformed cells, also have a high requirement for glutamine[8]. These amino acids represent the bare minimum and the majority of cells will grow better in the presence of additional amino acids. For example, the growth requirements of single cells are more extensive than for larger cell populations. Lockart and Eagle[9] found that seven other amino acids, especially serine, had to be included in the medium to support the growth of single cells. Only L-isomers of amino acids are used by cells, but the presence of excess quantities of the D-isomers of the 'essential' amino acids has no effect. However, D-isomers of some non-essential amino acids may be inhibitory[10]. Cells contaminated with mycoplasma have an abnormally high requirement for arginine[11] and this may possibly explain the relatively high concentration of arginine in some media.

The majority of essential vitamins appear to be of the B group. They include p-aminobenzoic acid, biotin, choline, folic acid, nicotinic acid, pantothenic acid, pyridoxal, riboflavin, thiamin and inositol. Most of these are known to form integral parts of coenzymes involved in metabolism and hence some media include the coenzymes as well. Media designed to be used in the absence of serum also often include the fat soluble vitamins. Another important group of substances found in complex media are the reducing agents. glutathione, ascorbic acid and L-cysteine. Cells will use nucleic acid derivatives preferentially rather than synthesising nucleic acids directly from simpler molecules and thus these, too, are common constituents of defined media. The role of serum in medium is the subject of many conflicting reports. What is clear, however, is that in spite of highly sophisticated attempts to develop a serum-free medium, very few cells can be persuaded to grow in the absence of serum. Even when this is achieved, growth is always markedly improved by its addition.

A final group of medium additives comprising the antibiotics plays no part in cell nutrition or metabolism. All media used in cell culture readily support the growth of microbial contaminants. Antibiotics provide a useful adjunct to careful sterile technique for the prevention of contamination. There is no reason why they should not be used with reasonable discrimination, providing the potential dangers are fully understood. The main dangers are suppression rather than elimination of contaminants, and the creation of a false sense of security by the constant addition of antibiotics. When they are used routinely for cell lines, they should be withdrawn at intervals in order to unmask possible low grade contamination. It must also be remembered that some antibiotics are quite toxic at their effective concentrations and that some, for example mycostatin and penicillin, are relatively unstable. Thus it is possible for these to have decayed and become ineffective before the end of an experiment.

The most commonly used antibiotics are penicillin and streptomycin, although dihydrostreptomycin is suggested as an alternative to the latter since it is as effective, but considerably less toxic at the same concentration[12]. Tomkins and Ferguson[13] proposed the following list as suitable for the elimination of most micro-organisms from cell cultures: kanamycin, neomycin, fungizone, chloramphenicol, viomycin and polymyxin B.

Mycoplasmas present a greater problem than most other organisms, because they are more difficult to detect and eliminate. Myo-crisin, which is slightly toxic, or tylosin tartrate[14], which can be used safely up to 100 µg/ml and is stable at +4°C, are used for the treatment of mycoplasmas. However, even these substances, in common with many tetracyclines tend to suppress rather than eradicate and should be used only when strictly necessary.

PRACTICAL CONSIDERATIONS

Many of the practical difficulties associated with the preparation of cell culture media have been alleviated by the appearance on the market of a wide range of fully tested media prepared from high quality reagents. For those wishing to undertake this work for themselves, the details of preparing complex media by means of stock concentrates of compatible components and the methods for obtaining various natural materials from living sources, have been fully described in a number of earlier treatises[15-17]. It is not proposed, therefore, to deal with this aspect, but rather consider some of the problems which still remain in association with commercial media.

Liquid medium presents few problems. It is generally supplied at working strength or as a tenfold concentrate. Working strength medium is usually supplied complete and only requires the addition of serum before use as a growth medium. Since it normally contains antibiotics, it is recommended that it be used fairly rapidly, and hence over a long period, many different batches would have to be used. Batches can vary slightly and thus it is impossible to guarantee exact replication of the conditions in each subsequent experiment. The same disadvantage applies to concentrated medium, but since sodium bicarbonate and the antibiotics are generally omitted, the medium can be used over longer periods. Complex defined media include a number of substances, for instance glutamine, which are relatively unstable in solution. These substances tend to be more stable at lower temperatures, hence medium should be kept at +4°C when not in use, or at −20°C for long-term storage.

As techniques in cell culture have improved, the need for highly standardised bulk media has increased. Great difficulties were experienced in propagating Hayflick's WI-38 cell line[18] when it was first distributed in Europe, and this was almost certainly due both to variations in purity of the components, and of the recipes followed in preparing the medium. Because of the limitations in the storage of liquid media, increasing attention has been paid to the preparation of media in a dry powdered form[19-21]. Dry medium, prepared either by ball-milling or by spray- or freeze-drying is now available from a number of commercial sources. Such media support the growth of cell cultures equally as well as media prepared in the traditional manner. Other advantages include lower costs, greater stability, uniformity of components by bulk buying of carefully selected ingredients and low transit costs. Although it is recommended that powdered medium is stored at $+4°C$, it is stable for long periods at room temperatures and hence may be sent long distances without harmful effects. Ball-milled media suffer from the slight disadvantage that component particles can assume an uneven distribution during long storage periods. For this reason it is best to purchase the medium in preweighed aliquots in amounts that satisfy monthly or weekly requirements. Some manufacturers provide the additional service of releasing samples of batches of dry media so that customers can ensure its suitability for their purposes before committing themselves to the purchase of large quantities.

One of the most crucial stages in the preparation of a medium is the sterilisation of the final product. Heat stable media may be sterilised by autoclaving. This method is both cheap and, providing due attention is paid to the conditions, highly effective. Autoclaving can result in changes in the medium, but these need not necessarily be deleterious. Sargeant and Smith[22] reported that a compound was formed, when glucose and phosphate ions were autoclaved together, which stimulated growth and the attachment of epithelial cells to glass. It was also found[23] that maintenance of cultures was impaired when the growth medium was filtered instead of autoclaved. For media with labile components, and this includes the majority of chemically defined or semidefined media, effective sterility is usually achieved by filtration. However, it has been shown[24] that Eagle's Media can be rendered thermostable at pH 4–4·5 by the incorporation of a succinate buffer. These autoclavable media require the addition of glutamine and bicarbonate solutions prior to use. Most filtration methods have their attendant problems. Ceramic candles were once widely used, but tended to impair the growth promoting potential. This was probably caused either by selective adsorption of components from the medium or by catalytic activity occuring on the porous surface of the candle. Seitz filtration suffers from the presence of a toxic substance in the asbestos pads which is extracted by the passage of medium[25]. It is generally agreed that membrane filtration represents the best method, although some membranes are supplied impregnated with a detergent to enable wetting and facilitate sterilisation, and this may be cytotoxic[26]. As with asbestos pads, toxicity can be removed by prewashing. It is important to avoid the use of chromium plated apparatus since copper may be leached out from the underlying brass by acid filtrates, giving a toxic product.

A wide range of sera is now available to enrich growth media. These are generally supplied untreated or inactivated by heating to $56°C$ for 30 min. Heat inactivation destroys complement and some contaminating viruses, thus rendering cultures grown on it more useful for virological procedures. Serum treated in this way may have slightly diminished growth promoting properties, but can be stored satisfactorily at $+4°C$. Unheated sera are rather less stable and should be stored at $-20°C$. In all cases where media or medium components are stored frozen, attention should be paid to the type of container and the provision of adequate space for expansion during freezing.

All media offered for sale are tested by the manufacturers for their ability to support growth and for freedom from toxicity. There is, however, some divergence of opinion on the best way of doing this. Medium designed for the growth of primary cultures is relatively easy to test. Cells which are sensitive to medium quality are used, and growth, morphology and evidence of toxicity in the test medium and in a standard control medium are carefully compared. The complexity of the test increases for medium designed for continuous cell culture. Two main possibilities exist, that is a plating efficiency test or a test involving a number of serial passages. Relative plating efficiency tests can certainly detect quite small differences in the quality of a medium, but it has been suggested[27] that high plating efficiency does not necessarily correlate with good growth promotion. This casts some doubt on the validity of plating efficiency as a method for assessing the growth potential of medium. Serial subcultivation tests are time consuming, but since the conditions of the test are the same as those under which the medium would normally be used, the results are probably more significant. It is generally found that if a medium will support the growth of a cell line for at least six subcultivations, thereafter the limit in the number of subcultivations, if any, is set by the cells themselves rather than by the medium. It is clear, therefore, that a medium should be tested over at least six serial subcultivations in order to be certain that it is suitable for the prolonged growth of cell lines.

BALANCED SALT SOLUTIONS

All balanced salt solutions (BSS) are developments from the physiological saline devised by Ringer[28]. His mixture consisted of the three cations, calcium, potassium and sodium in similar proportions to those existing in sea water or the blood of higher animals. Tyrode's solution[29] was the first BSS formulated specifically for supporting the metabolism of mammalian cells, but had the disadvantage that great care had to be taken during preparation to avoid calcium precipitation. Since then a number of BSS have appeared. None of these have any specific advantage so far as growth is concerned, but have been designed with improved buffering capacity or to prevent precipitation. The two most commonly used today are those of Earle[30] and Hanks[31]. Earle's BSS is more strongly buffered by virtue of a higher concentration of sodium bicarbonate. Hanks' BSS was primarily designed to equilibrate with air, rather than CO_2. It is particularly of use in conjunction with cells with which bicarbonate is toxic. Dulbecco and Vogt's phosphate buffered saline[32] is a valuable fluid for irrigating and washing cells and as a base for trypsin solutions. The composition of the BSS listed above is shown in *Table 1.1*. A more detailed account of the development of BSS and a description of lesser used specialist formulae is given in the review of Stewart and Kirk[33].

GROWTH MEDIA

NATURAL MEDIA

With the exception of serum, natural media are only rarely obtainable from commercial sources. This perhaps explains the current decline in their use for

Table 1.1 COMPOSITION OF BALANCED SALT SOLUTIONS (g/l)

	Phenol red	NaCl	KCl	CaCl$_2$ Anhydrous	MgCl$_2$· 6H$_2$O	MgSO$_4$· 7H$_2$O	Na$_2$HPO$_4$· H$_2$O	NaH$_2$PO$_4$· 2H$_2$O	KH$_2$PO$_4$	Glucose	NaHCO$_3$
Ringer	–	9·00	0·42	0·25	–	–	–	–	–	–	–
Tyrode	–	8·00	0·20	0·20	0·10	–	–	0·05	–	1·00	1·00
Earle	0·02	6·80	0·40	0·20	–	0·20	–	0·14	–	1·00	2·20
Hanks	0·02	8·00	0·40	0·14	–	0·20	0·06	–	0·06	1·00	0·35
Dulbecco and Vogt	0·02	8·00	0·20	0·10	0·10	–	1·42	–	0·20	–	–

until quite recently they were commonly employed. Thus in 1954, Malherbe[34] described a medium for the growth of human embryonic cells consisting of allantoic or amniotic fluid (90%) normal horse serum (5%) and bovine embryo extract (5%). In 1956 Jordan[35] used homologous human serum (35%), chick embryo extract (5%) and Hanks' BSS (60%) for the culture of human nasal cells. Plasma clot cultures[36], in which tissue is embedded in a coagulum caused by mixing chicken plasma and chick embryo extract, are still used when only small fragments of tissue are available to initiate a culture. It is possible to obtain freeze-dried or fresh chicken plasma in siliconised tubes for this purpose.

Apart from serum, the most frequently used medium ingredient from a natural source is lactalbumin hydrolysate. It is prepared by enzymic hydrolysis of a milk protein and is marketed by a number of companies. Melnick's medium[37] comprising Earle's BSS, lactalbumin hydrolysate and serum was originally designed for monkey kidney cell culture, but it provides an excellent growth medium for most primary cells. It must be pointed out, however, that commercial products can vary in colour, amino acid content and nutritional activity. It was found[38] that some cell lines lost their ability to attach to glass and that growth was poorer when batches were changed. The concentration of lactalbumin hydrolysate in the medium profoundly affects growth[39], with a sharp peak in optimum activity. As mentioned earlier[23], the method of sterilisation also markedly affects the performance of lactalbumin hydrolysate in the medium. A second protein digest which is a fairly frequent additive to culture medium, is tryptose phosphate broth[40]. It is particularly useful to enrich semi-defined growth media.

MEDIA COMPRISING BOTH NATURAL AND DEFINED COMPONENTS

Although natural media go part way to fulfilling the conditions described in the first paragraph of this chapter, they fall very far short of achieving the standardisation and control which is so important for current techniques. At the same time, completely defined media have two major defects. They support the growth of only a narrow range of highly adapted cells, and their performance is always improved by the addition of serum. Serum is, therefore, a constituent of the majority of growth media. Media have been designed in which only the components essential for cell growth have been added, but which require the addition of whole or dialysed serum. Probably the most frequently used media of this type were designed by Eagle[41, 42]. Eagle's media include the thirteen amino acids and seven vitamins found to be essential for growth dissolved in Earle's BSS (*Table 1.2*). A number of cells will grow in Eagle's medium supplemented with only albumin and inositol[43], although 5–10% serum is usually added.

Sera from a number of sources are used for cell culture. The most commonly used are bovine, but sera from horses, humans, rabbits and other sources are also used and can be obtained commercially. In general, the younger the animal, the better the growth provided by the serum. Hence serum from adult cattle tends to be poorer than that from young calves, whereas foetal calf serum is superior to both. In view of the tremendous importance of serum in cell culture medium, it is perhaps surprising that its function is not more fully understood. Opinions range over a number of possibilities. It seems almost certain that low molecular weight nutrient substances, found in association with, or absorbed to serum proteins play an important part. Dialysis can remove the growth promoting activity of serum[42, 44, 45], but this can be restored by the addition of proteose peptone (0·1%)[46]. Activity is not restored by adding certain vitamins,

Table 1.2 COMPOSITION OF EAGLE'S MEDIA IN mg/l
(mM equivalents for amino acids given in brackets)

	Basal Medium (1955)	Minimum Essential Medium (1959)
Amino acids		
L-arginine HCl	21 (0·1)	126 (0·6)
L-cystine	12 (0·05)	24 (0·1)
L-glutamine	292 (2·0)	292 (2·0)
L-histidine HCl	9·5 (0·05)	38 (0·2)
L-isoleucine	26 (0·2)	52 (0·4)
L-leucine	26 (0·2)	52 (0·4)
L-lysine HCl	36 (0·2)	72 (0·4)
L-methionine	7·5 (0·05)	15 (0·1)
L-phenyl alanine	18 (0·1)	32 (0·2)
L-threonine	24 (0·2)	48 (0·4)
L-tryptophane	4 (0·02)	10 (0·05)
L-tyrosine	18 (0·1)	36 (0·2)
L-valine	24 (0·2)	48 (0·4)
Vitamins		
Choline chloride	1	1
Folic acid	1	1
Inositol	2*	2
Nicotinamide	1	1
Calcium pantothenate	1	1
Pyridoxal HCl	1	1
Riboflavin	0·1	0·1
Thiamine	1	1
Salts (Earle's BSS)		
NaCl	6800	6800
KCl	400	400
$CaCl_2$	200	200
$MgSO_4.7H_2O$	200	200
$NaH_2PO_4.2H_2O$	150	150
$NaHCO_3$	2000	2000
Glucose	1000	1000

Biotin (1·0 mg/l is also often included.
* Not included in Eagle's original formula, but often a component of commercial media.

nucleic acid derivatives or amino acids. This indicates that peptides may be involved and this is further substantiated by Piez *et al.*[47]. There are indications, however, that the amino acids adsorbed to protein do play a part, since it has been found that serum at high concentrations may provide amino acids at concentrations which can reach threshold levels for survival in an otherwise deficient medium[48]. A further group of low molecular weight substances thought to be of importance are lipid sources[49-51].

High molecular weight fractions of serum are also thought to play a nutritive role[52]. Both albumin[53] and α and β globulins[54-56] have been implicated. However, at least one group claims that serum protein is not degraded or incorporated by cells[57]. High molecular weight fractions may, however, act in an entirely different manner. It is suggested that they may play a physical role, by protecting cells from mechanical damage[56, 58, 59]. or by acting as detoxifying agents[60]. Norkin et al.[51] showed that albumin was most effective in medium if the majority of low molecular weight substances bound to it, were extracted. This released binding sites which could then take up toxic substances from the medium. A further example of the physical action of serum is its ability to promote the attachment and spreading of cells to the substrate, particularly to glass[56, 61, 62]. This may be brought about by an alteration of the surface charge of the glass[63, 64]. The nature of the serum fraction which causes this is not clear, but is almost certainly associated with the α-globulins[65]. It has been identified as the glycoprotein fetuin[66], but another protein not identical to fetuin has also been found[67]. More recently, Wallis et al.[68] postulated that the function of serum in growth media was to inhibit proteolytic enzymes. They suggested that serum is required to inactivate residual trypsin remaining after enzymic disaggregation, together with proteases synthesised subsequently by the cells. It was concluded that 'toxic' sera may simply have little or no anti-enzyme component and that 'inactivation' may result in lowering its concentration and hence the growth promoting property of the serum. Which of these various functions serum fulfils, or the order of their importance, is still to be determined. The considerable complexity of serum makes a full understanding difficult, but until this is achieved, the discovery of a completely effective substitute will inevitably be delayed.

DEFINED MEDIA

From the discussion on the role of serum it can be seen that there are a number of possible approaches to the problem of serum substitution in media. Following the theory that serum was acting as a protective agent, Katsuta et al.[58] proposed alginic acid, dextran and polyvinylpyrrolidone (PVP) as alternatives. They adapted a strain of HeLa to grow on a medium containing 0·1% PVP (average molecular weight, 7×10^5), lactalbumin hydrolysate (0·4%) and yeast extract (0·8%) in a BSS[59]. Purified proteins, in particular insulin sometimes with the addition of inert polymers, have also been used[49, 69, 70]. Two of these groups[49, 70] include lipid sources in their medium and the former demonstrated that the methyl esters of long chain unsaturated fatty acids stimulated growth, whereas long chain saturated and branched chain fatty acids did not. Michl[71] worked for several years on a flattening fraction derived from serum. He later found that it could be replaced by carbamyl phosphate[72] which when added to a synthetic medium, provided a good substitute for whole serum, especially when insulin was present. Hams' medium F.10 (Table 1.3)[73] was specially devised to support the growth cells when supplemented with purified proteins. Each of the

Table 1.3 COMPOSITION OF HAM'S NUTRIENT MIXTURE, F.10 IN mg/l
(mM equivalents for amino acids given in brackets)

Amino Acids		Vitamins	
L-alanine	8·91 (0·1)	Biotin	0·02
L-arginine HCl	211·00 (1·0)	Calcium pantothenate	0·71
L-aspartic acid	13·30 (0·1)	Choline chloride	0·69
L-asparagine HCl	15·00 (0·1)	Folic acid	1·32
L-cysteine HCl	31·50 (0·2)	Inositol	0·54
L-glutamic acid	14·70 (0·1)	Nicotinamide	0·61
L-glutamine	146·20 (0·1)	Pyridoxine HCl	0·20
Glycine	7·51 (0·1)	Riboflavin	0·37
L-histidine HCl	21·00 (0·1)	Thiamine HCl	1·01
L-isoleucine	2·60 (0·02)	Thymidine	0·73
L-leucine	13·10 (0·1)	Vitamin B_{12}	1·36
L-lysine HCl	29·30 (0·1)		
L-methionine	4·48 (0·03)	Nucleic Acid Derivative	
L-phenyl alanine	4·96 (0·03)	Hypoxanthine	4·08
L-proline	11·50 (0·1)		
L-serine	10·50 (0·1)	Lipid Source	
L-threonine	3·57 (0·03)	Lipoic acid	0·20
L-tryptophane	0·60 (0·003)		
L-tyrosine	1·81 (0·01)	Carbohydrate source other than Glucose	
L-valine	3·50 (0·03)	Sodium pyruvate	110·0
		Salts	
		$FeSO_4 . 7H_2O$	1·52
		$CuSO_4 . 5H_2O$	0·004
		$ZnSO_4 . 7H_2O$	0·05
		NaCl	7400·00
		KCl	285·00
		Na_2HPO_4	153·60
		KH_2PO_4	83·00
		$MgSO_4 . 7H_2O$	152·80
		$CaCl_2$	33·30
		$NaHCO_3$	1200·00
		Glucose	1100·00

components was added at an experimentally determined optimum concentration. It was designed to be used either with serum albumin and fetuin or with low concentrations of serum.

For some time the goal of many cell biologists has been to produce a chemically defined cell culture medium free from added protein. As early as 1911 Lewis and Lewis[74] attempted to do this when they found that glucose or amino acids and peptides prolonged the survival of chick embryo tissues in saline solutions. Modern attempts to solve the problem stem from 1950 and the Medium 199 of Morgan, Morton and Parker[75] (*Table 1.4*). Medium 199 included almost all the amino acids, vitamins, several nucleic acid derivatives, accessory growth factors, lipid sources and Earle's BSS supplemented with ferric nitrate. This exceedingly complex medium is still widely used as a maintenance medium and when supplemented with serum, as a growth medium. In the complete absence of serum, however, it is only able to maintain cells for relatively short periods. The medium was later modified[76] to give Medium 858. This was considerably more efficient that Medium 199. In particular it restored

Table 1.4 COMPOSITION OF SYNTHETIC MIXTURE NO. 199 IN mg/l
(mM equivalents for amino acids are given in brackets)

Amino Acids		Coenzymes	
L-alanine*	25·0 (0·30)	A.T.P.	10·0
L-arginine HCl	70·0 (0·40)		
L-aspartic acid*	30·0 (0·25)		
L-cystine	20·0 (0·75)	Reducing Agents	
L-glutamic acid*	75·0 (1·00)	Ascorbic acid	0·05
L-glutamine	100·0 (0·07)	L-cysteine HCl	0·1
Glycine	50·0 (0·70)	Glutathione	0·05
L-histidine HCl	20·0 (0·10)		
L-hydroxyproline	10·0 (0·75)		
L-isoleucine*	20·0 (0·15)	Nucleic acid derivatives	
L-leucine*	60·0 (0·45)	Adenine	10·0
L-lysine	70·0 (0·20)	Guanine	0·3
L-methionine*	15·0 (0·20)	Hypoxanthine	0·3
L-phenylalanine*	25·0 (0·15)	Thymine	0·3
L-proline	40·0 (0·35)	Uracil	0·3
L-serine*	25·0 (0·25)	Xanthine	0·3
L-threonine*	30·0 (0·25)	Adenylic acid	0·2
L-tryptophane*	10·0 (0·005)		
L-tyrosine	40·0 (0·20)		
L-valine*	25·0 (0·20)	Lipid source	
		Cholesterol	0·2
		Tween 80	5·0
Vitamins			
p-aminobenzoic acid	0·05		
Biotin	0·01	Carbohydrate sources other than glucose	
Calcium pantothenate	0·01	2-deoxy-D-ribose	0·5
Choline chloride	0·5	D-ribose	0·5
Folic acid	0·01	Sodium acetate	50·0
Inositol	0·05		
Niacin	0·025		
Nicotinamide	0·025	Salts	
Pyridoxal HCl	0·025	NaCl	6800·0
Pyridoxine HCl	0·025	KCl	400·0
Riboflavin	0·01	CaCl$_2$	200·0
Thiamine HCl	0·01	MgSO$_4$. 7H$_2$O	200·0
Vitamin A	0·10	NaH$_2$PO$_4$	140·0
Vitamin D	0·10	NaHCO$_3$	2200·0
Vitamin E	0·01	Fe(NO$_3$)$_3$. 9H$_2$O	0·1
Vitamin K	0·01		
		Glucose	1000·0

* Double quantity of DL amino acid in original formula.

the abnormally high oxidation potential of Medium 199 to nearer physiological levels by large increases in the amounts of the reducing agents, L-cysteine HCl, ascorbic acid and glutathione. Five B vitamins, thiamine, riboflavin, niacin, niacinamide and pantothenate were omitted because they were constituents of six coenzymes which were added. Two years later, Medium 858 was modified to give CMRL-1066 Medium[77]. Four fat soluble vitamins, A, D, E, and K were replaced by the five B vitamins previously left out. This medium, the result of seven years of modifications and hundreds of experimental media, is capable of supporting a continuously cultured cell line in the absence of serum. A much simpler medium was later developed by Waymouth. Her Medium 752/1 (*Table 1.5*)[78] is now widely used. Over a long period of time a range of chemically

Table 1.5 COMPOSITION OF WAYMOUTH'S CHEMICALLY DEFINED MEDIUM, MB 752/1 IN mg/l
(mM equivalents of amino acids given in brackets)

Amino Acids		*Reducing Agents*	
L-arginine HCl	75·0 (3·6)	Ascorbic acid	17·5
L-aspartic acid	60·0 (4·6)	L-cysteine HCl	90·0
L-cystine	15·0 (0·06)	Glutathione	15·0
L-glutamic acid	150·0 (10·2)		
L-glutamine	350·0 (23·8)		
Glycine	50·0 (6·6)	*Nucleic Acid Derivatives*	
L-histidine	150·0 (8·0)	Hypoxanthine	25·0
L-isoleucine	25·0 (1·9)		
L-leucine	50·0 (3·8)		
L-lysine	240·0 (14·2)	*Salts*	
L-methionine	50·0 (3·4)	NaCl	6000·0
L-phenylalanine	50·0 (3·0)	KCl	150·0
L-proline	50·0 (4·4)	$CaCl_2 . 2H_2O$	120·0
L-threonine	75·0 (6·4)	$MgCl_2 . 6H_2O$	240·0
L-tryptophane	40·0 (2·0)	$MgSO_4 · 7H_2O$	200·0
L-tyrosine	40·0 (2·2)	Na_2HPO_4	300·0
L-valine	65·0 (5·5)	KH_2PO_4	80·0
		$NaHCO_3$	2240·0
Vitamins			
Biotin	0·02	Glucose	5000·0
Choline chloride	250·0		
Calcium pantothenate	1·0		
Folic acid	0·4		
Inositol	1·0		
Nicotinamide	1·0		
Pyridoxine HCl	1·0		
Riboflavin	1·0		
Thiamine HCl	10·0		
Vitamin B_{12}	0·2		

defined NCTC media have been developed in Dr. Earle's laboratory in Bethesda. One of these, NCTC 109 (*Table 1.6*)[79] supports the growth of a number of cell lines, including both malignant and normal cells of varying morphology and originating from mouse, hamster, human and monkey sources. The medium has been greatly simplified so that almost half the vitamins, all coenzymes, reducing substances and lipid sources have been omitted[80]. The simplified media supported the growth of many of the cell lines, but revealed in some cases a requirement for inositol and more rarely, for vitamin B_{12}. *Table 1.7* gives an outline comparison of the chemically defined media described above. Most of them are very complex, the total number of components ranging from 72 to 41. Nevertheless, some fifty years after the pioneer work of Lewis

Table 1.6 COMPOSITION OF NCTC 109 MEDIUM IN mg/l
(mM equivalents of amino acids given in brackets)

Amino Acids

L-alanine	31·48 (3.5)
L-α-amino-n-butyric acid	5·51 (0·5)
L-arginine HCl	31·00 (1·5)
L-asparagine	8·09 (0·6)
L-aspartic acid	9·91 (0·7)
L-cystine	10·49 (0·4)
D-glucosamine	3·20 (0·2)
L-glutamic acid	8·26 (0·5)
L-glutamine	135·73 (9·3)
Glycine	13·51 (1·8)
L-histidine	19·73 (1·3)
L-hydroxyproline	4·09 (0·3)
L-isoleucine	18·04 (1·4)
L-leucine	20·44 (1·6)
L-lysine	30·75 (2·1)
L-methionine	4·44 (0·3)
L-ornithine	7·38 (0·6)
L-phenylalanine	16·53 (1·0)
L-proline	6·13 (0·5)
L-serine	10·75 (1·0)
L-taurine	4·18 (0·3)
L-threonine	18·93 (1·6)
L-tryptophane	17·50 (0·9)
L-tyrosine	16·44 (0·9)
L-valine	25·00 (2·1)

Vitamins

p-aminobenzoic acid	0·125
Biotin	0·025
Calcium pantothenate	0·025
Choline chloride	1·25
Folic acid	0·025
Inositol	0·125
Niacin	0·0625
Nicotinamide	0·0625
Pyridoxal HCl	0·0625
Pyridoxine HCl	0·0625
Riboflavin	0·025
Thiamine HCl	0·025
Vitamin A	0·025
Vitamin D	0·25
Vitamin E	0·025
Vitamin K	0·25
Vitamin B_{12}	10·00

Coenzymes

DPN	7·0
TPN	1·0
Coenzyme A	2·5
Cocarboxylase	1·0
Flavin adenine dinucleotide	1·0
UTP	1·0

Reducing Agents

Ascorbic acid	49·9
L-cysteine HCl	259·9
Glutathione	10·1

Nucleic acid derivatives

Deoxyadenosine	10·0
Deoxycytidine	10·0
Deoxyguanosine HCl	10·0
Thymidine	10·0
5-methyl cytosine	0·1

Lipid source

Tween 80	22·5

Carbohydrate sources other than glucose

Glucuronolactone	1·8
Na-glucuronate H_2O	1·8
Sodium acetate . $3H_2O$	50·0

Salts

NaCl	6800·0
KCl	400·0
$CaCl_2$	200·0
$MgSO_4$. $7H_2O$	200·0
Na_2HPO_4	140·0
$NaHCO_3$	2200·0
Glucose	1000·0

Table 1.7 OUTLINE COMPARISON OF SOME CHEMICALLY DEFINED MEDIA

	107	109	NCTC Media (1956–1964)					199 (1950)	858 (1955)	1066 (1957)	752/1 (1959)
			117	126	131	132	133				
Amino acids	25	25	25	25	25	25	25	20	21	20	17
Vitamins	17	17	17	8	9	9	10	16	11	12	10
Coenzymes	6	6	–	–	–	–	–	1	6	6	–
Reducing agents	3	3	3	2	2	2	2	3	3	3	3
Nucleic acid derivatives	5	5	2	2	2	2	2	7	1	1	1
Lipid sources	5	1	1	1	1	1	1	2	2	2	–
Carbohydrates other than glucose	2	2	2	1	1	1	1	2	1	1	–
Sodium acetate	+	+	–	+	+	+	+	+	+	+	–
Ethyl alcohol	+	+	+	+	–	–	–	–	–	–	–
B.S.S.	++	++	++	++	++	++	++	++	++	++	–
Glucose	+	+	+	+	+	+	+	+	+	+	++

and Lewis[74]. growth of cells in completely defined medium has finally been achieved. In no other field of culture is purity of components and meticulous preparation and storage of greater importance.

MAINTENANCE MEDIA

The essential feature of a maintenance medium is that it should preserve the cell culture in good condition for several weeks, without impairing any of the properties required for experimental purposes. Growth, which may cause the cell sheet to detach from the glass or make microscopic examination difficult, should be minimised, and there should be no reduction in viral sensitivity[81]. In many cases it is sufficient merely to reduce the serum concentration or to eliminate it altogether from the medium used for growth. Serum has the disadvantage that it may encourage cell growth or contain viral inhibitors[82]. Most of the chemically defined media described earlier, although unable to support the growth of the majority of cells in the absence of serum, make excellent maintenance media. Medium 199 is still extensively used as a maintenance medium for primary cell cultures during viral vaccine manufacture. Several media have been developed purely for the purpose of culture maintenance. Most of them have been designed in order to eliminate serum, and have included gelatin[83] and skim milk[84]. One of the most effective was devised by Smith[81], whose medium contained liver digest ultrafiltrate.

Maintenance is not simply a matter of finding a good medium. There is evidence that the potential for long maintenance is determined early in the history of a culture. Both the methods of cell dispersal and the growth conditions markedly effect maintenance. By the time a culture has become confluent, little can be done to prolong maintenance beyond the limit fixed in the growth period[23].

CHOICE OF MEDIUM

Many of the companies marketing cell culture media now include a considerable range in their catalogues. In only a few cases, are any indications of their use given. Many media were designed with a specific cell line or cell type in mind, but they can normally be used for a much wider range of cultures. Established cell lines are often adapted to grow on a particular medium, and unless it has components that might interfere with the work planned, this should be the medium of choice. It can however be difficult to decide, which of the many media to choose when setting up new cultures or attempting to establish new cell lines.

For large scale primary cell cultures, lactalbumin hydrolysate with serum (10%) is simple, cheap and highly efficient. The use of Medium 199 and its derivatives enables the serum concentration to be reduced while still giving good growth. Ham's medium F.10 with 2–10% serum[73] is another excellent growth medium for primary cultures and also for cell lines. A later modification of this medium, Ham's F.12, was specifically designed to support single cell growth[85]. Eagle's media are the most commonly used for cell line culture. The simplest 'Basal' formulation[41] is used for many diploid cell cultures although the amounts of amino acids in this media may be limiting[86]. For this reason medium with double the concentration of amino acids and vitamins[87] is often preferred. Modified Eagle's medium with triple-strength amino acids is also available from some sources. Eagle's Minimum Essential Medium[42] is generally used for established cell lines, although the amounts of arginine and histidine may be somewhat excessive. McCoy's 5a medium[88] is particularly suitable for the most fastidious cell lines as well as for the growth of primary cell cultures, cells from tissue biopsies and leucocytes, although many workers prefer RPMI 1640 for the latter[89].

It is perhaps surprising just how readily mammalian cells will survive and grow on such a wide variety of media. Difficulties can sometimes occur in adapting cell lines to different types of related media or even to different batches of the same medium. Failure to describe the original medium accurately when they were first published, is sometimes at the root of the problem. It is not clear from some descriptions of media, whether hydrated or anhydrous salts were used, or if the amino acids were in the L or DL form. It is thus possible to make up two batches of the same medium with different tonicities and varying properties. It is of fundamental importance, therefore, as media increase in complexity, to describe each component accurately and to give the source from which it was obtained. Only by doing this can any degree of standardisation be obtained and the results from different laboratories properly assessed.

REFERENCES

1. GOMORI, G., *Proc. Soc. exp. Biol. Med.*, **62**, 33 (1946)
2. SWIM, H. E. and PARKER, R. F., *Science, N.Y.*, **122**, 466 (1955)
3. MARTIN, G. M., *Proc. Soc. exp. Biol. Med.*, **116**, 167 (1964)
4. SHIPMAN, C. JR., *Proc. Soc. exp. Biol. Med.*, **130**, 305 (1969)
5. LEIBOVITZ, A., *Am. J. Hyg.*, **78**, 173 (1963)
6. PHILLIPS, H. J. and ANDREWS, R. V., *Expl Cell Res.*, **16**, 678 (1959)
7. EAGLE, H., *J. biol. Chem.*, **214**, 839 (1955)
8. CHANG, R. S., *J. exp. Med.*, **113**, 405 (1961)
9. LOCKART, R. Z. JR. and EAGLE, H., *Science, N.Y.*, **129**, 252 (1959)
10. BEST, N. H., MILLS, S., LEACH, K. K., DURHAM, N. N. and LEACH, F. R., *Expl Cell Res.*, **31**, 13 (1963)
11. SCHIMKE, R. T. and BARILE, M. F., *J. Bact.*, **86**, 195 (1963)
12. MOSKOWITZ, M. and KELKER, N. E., *Science, N.Y.*, **141**, 647 (1963)
13. TOMKINS, G. A. and FERGUSON, J., *Aust. J. exp. Biol. med. Sci.*, **43**, 743 (1965)
14. CROSS, G. F., GOODMAN, M. R. and SHAW, E. J., *Aust. J. exp. Biol. med. Sci.*, **45**, 201 (1967)
15. PARKER, R. C., *Methods of Tissue Culture*, 3rd edn, Pitman Medical, London (1961)
16. PAUL, J., *Cell and Tissue Culture*, 4th edn, Livingstone, Edinburgh (1970)
17. PENSO, G. and BALDUCCI, D., *Tissue Cultures in Biological Research*, Elsevier, Amsterdam (1963)
18. HAYFLICK, L. and MOORHEAD, P. S., *Expl Cell Res.*, **25**, 585 (1961)
19. SWIM, H. E. and PARKER, R. F., *J. Lab. clin. Med.*, **52**, 309 (1958)
20. HAYFLICK, L., JACOBS, P. and PERKINS, F., *Nature, Lond.*, **204**, 146 (1964)
21. GREEN, A. E., SILVER, R. K., KRUG, M. D. and CORIELL, L. L., *Proc. Soc. exp. Biol. Med.*, **118**, 122 (1965)
22. SARGEANT, T. P. and SMITH, S., *Science, N.Y.*, **131**, 606 (1960)
23. WHITAKER, A. M., *Br. J. exp. Path.*, **50**, 347 (1969)

24. YAMANE, I., MATSUYA, Y. and JIMBO, K., *Proc. Soc. exp. Biol. Med.*, **127**, 335 (1968)
25. HOUSE, W., *Nature, Lond.*, **201**, 1242 (1964)
26. CAHN, R. D., *Science, N.Y.*, **155**, 195 (1967)
27. MOORHEAD, P. S., *Personal communication*
28. RINGER, S., *J. Physiol., Lond.*, **18**, 425 (1895)
29. TYRODE, M. V., *Archs Int. Pharmacodyn. Ther.*, **22**, 205 (1910)
30. EARLE, W. R., *Arch. exp. Zellforsch.*, **16**, 116 (1934)
31. HANKS, J. H., *J. cell. comp. Physiol.*, **31**, 235 (1948)
32. DULBECCO, R. and VOGT, M., *J. exp. Med.*, **99**, 167 (1954)
33. STEWART, D. C. and KIRK, P. L., *Biol. Rev.*, **29**, 119 (1954)
34. MALHERBE, H., *Proc. Soc. exp. Biol. Med.*, **86**, 124 (1954)
35. JORDAN, W. S. JR., *Proc. Soc. exp. Biol. Med.*, **92**, 867 (1956)
36. CARREL, A., *J. exp. Med.*, **38**, 407 (1923)
37. MELNICK, J. L., *Ann. N. Y. Acad. Sci.*, **61**, (4), 754 (1955)
38. TAKAOKA, T., KATSUTA, H., KANEKO, K., KAWANA, M. and FURUKAWA, T., *Jap. J. exp. Med.*, **30**, 393 (1960)
39. FLOSS, D. R., *Expl Cell Res.*, **34**, 603 (1964)
40. GINSBERG, H. S., GOLD, E. and JORDAN, W. S. JR., *Proc. Soc. exp. Biol. Med.*, **89**, 66 (1955)
41. EAGLE, H., *Science, N.Y.*, **122** 501 (1955)
42. EAGLE, H., *Science, N.Y.*, **130**, 432 (1959)
43. CHANG, R. S. and GEYER, R. P., *J. Immun.*, **79**, 455 (1957)
44. EAGLE, H., *Proc. natn. Acad. Sci. U.S.A.*, **46**, 427 (1960)
45. METZGAR, D. P. JR. and MOSKOWITZ, M., *Science, N.Y.*, **130**, 1421 (1959)
46. HARRIS, M., *Proc. Soc. exp. Biol. Med.*, **102**, 468 (1959)
47. PIEZ, K. A., OYAMA, V. I., LEVINTOW, L. and EAGLE, H., *Nature, Lond.*, **188**, 59 (1960)
48. DUPREE, L. T., SANDFORD, K. K., WESTFALL, B. H. and COVALESKY, A. B., *Expl Cell Res.*, **28**, 381 (1962)
49. TYTELL, A. A. and NEUMAN, R. E., *Fedn Proc. Fedn Am. Socs exp. Biol.*, **19**, 385 (1960)
50. HAM, R. G., *Science, N.Y.*, **140**, 802 (1963)
51. NORKIN, S. A., CZERNOBILSKY, B., GRIFFITH, E. and DUBIN, I. N., *Archs Path.*, **80**, 273 (1965)
52. FRANCIS, M. D. and WINNICK, T., *J. biol. Chem.*, **202**, 273 (1953)
53. TOZER, B. T. and PIRT, S. J., *Nature, Lond.*, **201**, 375 (1964)
54. KENT, H. N. and GEY, G. O., *Proc. Soc. exp. Biol. Med.*, **94**, 205 (1957)
55. KENT, H. N. and GEY, G. O., *Science, N.Y.*, **131**, 666 (1960)
56. HOLMES, R. and WOLFE, S. W., *J. biophys. biochem. Cytol.*, **10**, 389 (1961)
57. KING, D. W., BENSCH, K. G. and HILL, R. B. JR., *Science, N.Y.*, **131**, 106 (1960)
58. KATSUTA, H., TAKAOKA, T., HOSAKA, S., HIBINO, M., OTSUKI, I., HATTORI, K., SUZUKI, S. and MITAMURA, K., *Jap. J. exp. Med.*, **29**, 45 (1959)
59. KATSUTA, H., TAKAOKA, T., FURUKAWA, T. and KAWANA, M., *Jap. J. exp. Med.*, **30**, 147 (1960)
60. MOSKOWITZ, M., SCHENCK, D. M. and AMBORSKI, G., *Fedn Proc. Fedn Am. Socs exp. Biol.*, **22**, 383 (1963)
61. LIEBERMAN, I. and OVE, P., *J. biol. Chem.*, **233**, 637 (1958)
62. NORDLING, S., SAXEN, E. and PENTTINEN, K., *Acta path. microbiol. scand.*, **63**, 28 (1965)
63. RAPPAPORT, C. and BISHOP, C. B., *Expl Cell Res.*, **20**, 580 (1960)
64. NORDLING, S., PENTTINEN, K. and SAXEN, E., *Expl Cell Res.*, **37**, 161 (1965)
65. WEISS, L., *Expl Cell Res.*, **17**, 499 (1959)
66. FISHER, H. W., PUCK, T. T. and SATO, G., *Proc. natn Acad. Sci. U.S.A.*, **44**, 4 (1958)
67. LIEBERMAN, I. and OVE, P., *J. biol. Chem.*, **234**, 2754 (1959)
68. WALLIS, C., VER, B. and MELNICK, J. L., *Expl Cell Res.*, **58**, 271 (1969)
69. NAGLE, S. C. JR., TRIBBLE, H. R. JR., ANDERSON, R. E. and GARY, N. D., *Proc. Soc. exp. Biol. Med.*, **112**, 340 (1963)
70. HIGUCHI, K., *J. infect. Dis.*, **112**, 213 (1963)
71. MICHL, J., *Expl Cell Res.*, **23**, 324 (1961)
72. MICHL, J., *Nature, Lond.*, **207**, 412 (1965)
73. HAM, R. G., *Expl Cell Res.*, **29**, 515 (1963)
74. LEWIS, M. R. and LEWIS, W. H., *Anat. Rec.*, **5**, 277 (1911)
75. MORGAN, J. F., MORTON, H. J. and PARKER, R. C., *Proc. Soc. exp. Biol. Med.*, **73**, 1 (1950)
76. HEALY, G. M., FISHER, D. C. and PARKER, R. C., *Proc. Soc. exp. Biol. Med.*, **89**, 71 (1955)

15 718121

MAMMALIAN CELL CULTURE MEDIA   19

77. PARKER, R. C., CASTOR, L. N. and McCULLOCH, E. A., *Spec. Publs. N.Y. Acad. Sci.*, **5,** 303 (1957)
78. WAYMOUTH, C., *J. natn Cancer Inst.*, **22,** 1003 (1959)
79. EVANS, V. J., BRYANT, J. C., FIORAMONTI, M. C., McQUILKIN, W. T., SANFORD, K. K. and EARLE, W. R., *Cancer Res.*, **16,** 77 (1956)
80. EVANS, V. J., BRYANT, J. C., KERR, H. A. and SCHILLING, E. L., *Expl Cell Res.*, **36,** 439 (1964)
81. SMITH, S. E., *Br. J. exp. Path.*, **42,** 232 (1961)
82. TOMLINSON, A. J. H., DAVIES, J. R. and MADIGAN, E. M., *Mon. Bull. Minist. Hlth.*, **23,** 116 (1964)
83. FRIEDMAN, M. and LIEBERMAN, M., *Fedn Proc. Fedn Am. Socs exp. Biol.* **15,** 587 (1956)
84. BARON, S. and LOW, R. J., *Science, N.Y.* **128,** 89 (1958)
85. HAM, R. G., *Proc. natn. Acad. Sci. U.S.A.*, **53,** 288 (1965)
86. GRIFFITHS, J. B. and PIRT, S. J., *Proc. R. Soc.*, **168,** 421 (1967)
87. MACPHERSON, I. and STOKER, M., *Virology, 16,* 147 (1962)
88. McCOY, T. A., MAXWELL, M. and KRUSE, P. F., *Proc. Soc. exp. Biol. Med.*, **100,** 115 (1959)
89. MOORE, G. E., GERNER, R. E. and FRANKLIN, H. A., *J. Am. med. Ass.*, **199,** 87 (1967)

A/599.08

TWO

Sterile filtration

JACK G. MULVANY

Millipore (U.K.) Ltd., Heron House, Wembley, Middlesex

INTRODUCTION

The art of filtration is one of the oldest of chemical processing techniques. Early records of the major cultures indicate that filtration has long been a known and practised art. Through the centuries, men have used such materials as sand, clays, charcoal, string, cloth, ground coal, paper, felt, mesh and various powders along with a wondrous assortment of apparatus designed for filtration. Some techniques made use of pots loaded with filter media. Others, still in use, employed presses designed to hold sheets, centrifuges containing filter elements, rotating drums and large percolation beds.

The degree of filtration required varies widely in modern processes. The nature of the fluid, the flow rates involved, the level of particulate and microbiological removal required, all figure importantly in the choice of a filtration process. Sterilising filtration is just one example of filtration practice, however, when the aim of filtration is sterility, even 99·995% efficiency is not good enough. Every viable organism in the entire batch being filtered must be removed. To do this therefore it is necessary to employ those filtration techniques where the absolute removal of micron and sub-micron particles can be achieved. Obviously in this critical application the selection of the filter is extremely important and a knowledge of the characteristics of various types of filtration media is obviously of great advantage. At first sight, this latter task may seem impossible for those people to whom filtration is just one of the many services on which they rely to perform their basic function. However, this task is not so difficult as it seems because the seemingly bewildering variety and number of filter materials in use can all be grouped, rather simply, into two general categories: namely depth and screen filters.

In modern biological techniques the science of sterilising filtration is of fundamental importance as in a large number of cases it provides the only way of sterilising many fluids. Biological fluids such as tissue culture media cannot be heat sterilised as it would cause the breakdown of prime components of the media, such as the serum fraction. For obvious reasons vaccines and also many antibiotics have to be filter-sterilised, as in the case of the latter, many are heat labile. There are a host of other liquids which can only be sterilised by filtration and these are obviously far too numerous to detail here. The three main areas which require most experience and care when performing sterilising filtration are the selection of the filter, the design of the filtration system and finally, the type of pre-treatment which is employed. All of these major areas will be fully discussed in this chapter.

CHARACTERISTICS OF FILTRATION MEDIA

(i) DEPTH FILTRATION

Depth filters usually consist of fibres, particles or fragmented material of some sort, that has been pressed or bonded together to form a tortuous maze of flow passages (Fig. 2.1). Although the pores of a depth filter can be made coarse or

Fig. 2.1. Depth filters are generally constructed by compressing or bonding together particles, fibres or matted fragments of some suitable material to form tortuous flow passages in which the contaminants will become trapped

fine, depending on the degree of compaction or the size of the fibres, the flow channels are inherently random and variable in size. The depth filter medium collects particles or micro-organisms within its matrix, because its pores are generally much larger than the particles being collected. The particles move through the tortuous pores until they contact the medium when they are retained by a physical attraction, by gravity, or by becoming lodged in crevices (Fig. 2.2). Particle retention generally is controlled by changing the density of the structure with pressure, heat or sometimes by impregnating fibres with resinous compounds. The usual method of specifying a depth filter is an arbitrary one based wholly on experience or by means of tests after fabrication.

Examples of sterilising-grade filters are asbestos pads, ceramic candles, porous sintered glass, or stainless steel, and complex laminates of asbestos, glass and paper. Because particles and micro-organisms become trapped throughout the matrix of a depth filter, its dirt-holding capacity is relatively high. This advantage manifests itself in a long life or in the ability to handle solutions that are highly contaminated.

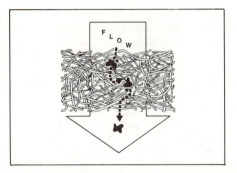

Fig. 2.2. Particles entering a depth filter stop at the point where resistance encountered is equal to the driving force. Any subsequent increase in pressure will drive them deeper, sometimes through the filter

Depth filters have several inherent limitations, however, that have an important bearing on sterile-filtration applications. For instance, when a particle contained in a fluid wends its way through a depth filter, it follows the path of least resistance until it adheres at some point or becomes wedged into place. When pressure differential increases, which often happens as a result of filter clogging or an increase in operating pressure, particles and micro-organisms are driven deeper into the matrix. Eventually, some will work through and into the filtrate. When a particle passes through, the result is a loss in efficiency and a corresponding drop in the quality of the filtrate. When a viable organism gets through, sterility is lost. For this reason, sterilising depth filters are generally limited to a maximum pressure differential of 15–20 p.s.i. Depth filters are given nominal rather than absolute ratings which are closely linked to allowable pressure differential. This type of filter medium does not remove all of the particles of a particular size from the fluid, and it is commonly designated as 90, 95 or 99% efficient at removing a particular sized test particle or micro-organism[1].

Grow-through of organisms trapped in the filter matrix is also a problem. If the conditions are right, these organisms will reproduce within the filter and successive generations will penetrate further into the filter, eventually emerging on the downstream side to contaminate the filtrate. This is most likely to occur during lengthy filtration runs. *Table 2.1* shows typical flow rates that are possible when using sterilising grades of asbestos pad filters[2] and it should be noted that the maximum pressure differential that is recommended when using these filters is approximately 15 p.s.i. One point of extreme importance before using a depth filter medium is to establish that the particular medium being used will have no detrimental chemical effect on the solution being filtered. The following two examples will serve to illustrate this point. As mentioned above many depth filter media are impregnated with resinous compounds to vary the particle retention capability. Certain of these resinous compounds may well be toxic to many biological systems and during the filtration run it is possible that they may be leached by the solution from the filter medium.

Table 2.1   TYPICAL FLOW RATES OF STERILISING GRADES OF ASBESTOS PAD FILTERS (Carlson–Ford)

Grade	Average density	Permeability	Mean pore size	Water output $(lh^{-1}m^{-2})$	Sugar syrup output $(lh^{-1}m^{-2})$
EKS2	0·352	0·00744	10·0	560	13·0
EKS	0·344	0·00713	13·7	750	17·5
EK	0·322	0·01116	15·4	1100	26·5
S10	0·320	0·01069	18·5	1400	34·0
S9	0·294	0·02433	27·3	2150	53·5

Secondly, it is well known that a sterilising grade of depth filter media is highly efficient at removing very small particles from the solution being processed. When sterilising a vaccine, it is important to remove any bacteria from the preparation, however, it is equally important to achieve a good titre of the filtrate; and this may not be possible when using a depth filter as a high percentage of the antigen will be removed in the matrix of the filter. Special precautions can be taken to reduce this effect, however, it cannot be completely eliminated. A severe disadvantage of all depth filters is their tendency to migrate—media migration. Fragments or fibres will slough off during filtration, this being more pronounced when there is a hydraulic surge or continued flexing of the filter matrix, thus contaminating the filtrate with fibres. Whilst in most cases this is not of much importance, where the solutions are being used for injection into humans or animals, special care must be taken because of the damaging effect of these fibres and particles when injected into the venous system[3].

(ii) SCREEN FILTRATION

Screen filters differ fundamentally from depth filters in their retention of particles. They are characterised by highly uniform, regularly spaced holes with sharp cut-offs in particle retention; that is, ratings are absolute because no particles whose smallest dimension exceeds the pore size of the screen can possibly pass through. The simplest examples are, window screens or per-forated sheets of metal or plastic (Fig. 2.3). Pore structure is pre-defined and retention is wholly confined to the surface, so dirt holding capacity is somewhat lower than that of a depth filter. Screen filter media are not prone to media migration because they are homogeneous and continuous. Thus sterilising filtration can be guaranteed by using a screen filter whose largest passage is smaller than the smallest know bacteria. Using specialised techniques screens may be fabricated from a wide range of materials such as stainless steel, nickel, nylon, polyester etc. Twill weaves may be manufactured so that the uniform

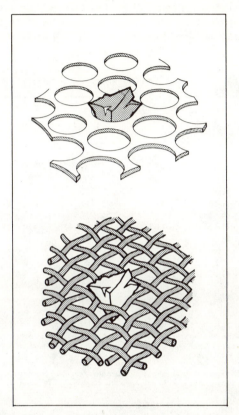

Fig. 2.3. Screen filters consist typically of continuous sheets of such materials as metal, plastic or woven strands of wire or textile filaments with highly uniform, regularly spaced pores

openings (pore size) may be as small as 2μm. Of course, this type of screen filter would not be useful in sterilising filtration as bacteria may be as small as 0·25μm.

(iii) MEMBRANE FILTERS

In the past fifteen years, membrane filters have been successfully developed to provide screen filters with openings of discrete sizes in the range 10μm down to 0·025μm. One of the processes which has gained more than any other from the development of membrane filters is sterilising filtration. For the first time, a screen filter of bacterial and even viral dimensions was available. Membrane filters, as manufactured by Millipore, are composed of thin porous sheets of pure, cellulose esters. Millipore filters are currently produced in twelve distinct pore sizes which are shown in *Table 2.2* along with their flow-rate data for both water and air. These unique filters are also produced from other polymers, such as nylon, polyvinyl chloride and PTFE. All of these filters are highly porous, with as much as 84% of their surface area taken up by pores. Ordinary woven screen filters by comparison have less than 50% of open area. The cellulose

Table 2.2   AIR AND WATER FLOW RATES FOR MF-MILLIPORE FILTERS

MF type	Mean pore size (μm)	Flow rate Water*	Air†
SC	8·0	850	55
SM	5·0	540	35
SS	3·0	400	20
RA	1·2	300	15
AA	0·8	212	11
(black)		212	11
DA	0·65	150	10
HA	0·45	64	4·5
(black)		64	4·5
PH	0·30	40	3·7
GS	0·22	21	2·5
VC	0·10	2	0·49
VM	0·05	1	0·31
VF	0·01	0·2	0·22

* Water flow rates are given in approximate milliliters per minute per square centimeter of filter area at 25°C with a pressure differential of 70 centimeters of mercury (13·5 p.s.i.).

† Air flow rates are given in approximate liters per minute per square centimeter of filter area at 25°C with a pressure differential of 70 centimeters of mercury (13·5 p.s.i.).

esters used in the manufacture of these filters are biologically inert and will not cause tissue reactions during implantation in animals and humans. Extensive tests by independent laboratories have shown that the Millipore membrane filters are non-pyrogenic and non-toxic. Perspex rings enclosed on each side by 14 mm membrane filter discs are routinely used for *in vivo* tissue implants and cellular studies in animals. Membrane filters are surface retentive, something which is true of all screen filters, so as mentioned above the dirt-handling capacity is lower than a depth filter approaching the same efficiency. It is usually necessary to precede a membrane filter with a suitable depth prefilter if a large volume of fluid or a highly contaminated fluid is to be processed. For analytical purposes, surface retention is a valuable characteristic because it enables the analytical microbiologist to concentrate all the particles or micro-organisms from a sample onto a planar surface where they are readily counted, sized, analysed or observed under a microscope. Properly supported, the membrane filter material will withstand a differential pressure of at least 10 000 p.s.i. The practical limits are usually imposed by the filter holder, rather than the filter. This characteristic of screen filters vividly highlights their difference to depth filters which, as mentioned above, are generally limited to approximately 15 p.s.i. in sterile filtration applications for the fear of breakthrough of micro-organisms.

Membrane filters are very thin (about 150 μm) so little liquid is ever retained within the filter matrix. This can be a major consideration if the fluid to be filtered happens to be costly and/or precious. Experience has shown that for volumes ranging from a few millilitres to even several thousand litres, filtration through Millipore filters and apparatus results in negligible losses. Adsorption from a fluid preparation by a membrane filter is generally negligible. Potency or titre are unaffected. The few exceptions are related primarily to virological applications where the virus particle in question is of a size of 0·1 μm and larger. When processing vaccines with small virus particles, such as polio vaccine, titre losses are negligible. Membrane filters are stable when dry to temperatures as high as approximately 200°C, they therefore may be autoclaved quite freely prior to being used for sterile filtration. The membrane filter's ability to withstand high pressures, coupled with this extraordinary porosity (hence, low resistance to flow), can be used to achieve high flow rates and hence much shorter filtration runs than with equal areas of depth filter. *Table 2.2* gives details of possible water flow rates using membrane filters and if these are compared to the flow rates that may be achieved with depth filters (*Table 2.1*) it can be readily seen that the flow rates with screen filters are up to forty times faster than those with depth filters under the same conditions. For highly viscous solutions, a filtration that may have taken two full days with a depth fil-ter can be accomplished by membrane filtration within a couple of hours—and with a filter unit that is many times smaller. Grow-through which is a common problem experienced with depth filters is, of course, non-existent with mem-brane filtration. Organisms cannot grow through the pores of a sterilising grade filter.

For filtering tissue culture media used to culture particularly fastidious cells, Millipore-MF filters may be obtained, on special request, from which the small amount of wetting agent normally used has been omitted.

(iv) MEMBRANE FILTER QUALITY CONTROL PROCEDURES

Millipore filters are subjected to rigorous testing and inspection procedures to ensure the utmost in high standard of quality. Among specific tests performed regularly are:

(*a*) Maximum pore size determination (bubble test). Bubble point testing is an extremely important control procedure which can also be employed by the user of these filters. For this reason the techniques of carrying out a bubble test have been fully explained later in this chapter.

(*b*) Flow rate determination.

(*c*) Determination of pore size distribution by means of mercury intrusion.

(*d*) Bacterial growth determination—organisms cultured on the filter surface must compare favourably with an agar plate control for the number recovered. colony formation and pigment production.

(*e*) Passage testing (sterile filtration test)—for the type GS (0·22 μm) sterilising filters a suspension of a *Pseudomonas* species with a minimum diameter of approximately 0·3 μm is filtered into sterile flasks through the GS test filter and an HA (0·45 μm) control filter. Both flasks are then incubated for 15 days at 37°C. The HA filtrate will become turbid, indicating *Pseudomonas* growth, but the GS filtrate should remain clear. If it does not, the entire production lot of GS filters is rejected.

In a recent study[4] to determine the effectiveness of sterilising filtration, it was found that the only system to give consistent sterile filtrates employed a Millipore filter and associated apparatus. Each filtration system was evaluated by challenging it with a heavy bacterial suspension of *Serratia marcescens*. The three other filtration media used in the evaluation programme were: diatomaceous earth (kieselguhr), sintered glass and asbestos pads. Thus it can be seen that the selection of the filter used for sterilising filtration is of prime importance. It is now widely accepted that only a membrane filter (Millipore) can consistently guarantee a sterile filtration under all conditions; however. due to its low dirt-handling capacity, it is often necessary to use a depth filter as a pretreatment to produce a viable and economic filtration system. *Table 2.3*

Table 2.3   CHEMICAL COMPATIBILITY OF MILLIPORE FILTERS

Chemical Resistance Table
(Static Conditions)

		MF Milli-pore	*Solv-inert**			*MF* Milli-pore	*Solv-inert**
Hydrocarbons	Hexane	I	I		Propylene Glycol	II	I
	Petroleum Ether	I	I		Ethylene Glycol	IV	I
	Pentane	I	I		Polyethylene Glycol	IV	I
	Benzene	I	I		Diethyl Acetamide	IV	II
	Toluene	I	I		Dimethyl Formamide	IV	II
	Xylene	I	I		Formaldehyde 37%	II	I
					Pyridine	IV	II
					Nitrobenzene	IV	II
Halogenated Hydrocarbons	Methyl Chloroform	I	I	**Miscellaneous Organic Compounds**	Methanol/ Chloroform (1–1)	II	I
	Chloroform	I	I		Triethyl Amine	III	I
	Carbon Tetrachloride	I	I		Hexamethylene- diamine	IV	I
	Methylene Chloride	III	II		Aniline	II	I
	Ethylene Chloride	II	II		Butyraldehyde	IV	I
	Trichloroethylene	II	I		Dimethyl Sulphoxide	IV	II
	Perchloroethylene	I	I		Tetrahydrofuran	IV	II
	Trichloroethane	I	I		Nitropropane	IV	I
					Phenol	III	I

	MF Milli-pore	Solv-inert*		MF Milli-pore	Solv-inert*
Alcohols			**Miscellaneous Solvents**		
Methyl Alcohol	IV	I	Mineral Spirits	I	I
Ethyl Alcohol	III	I	Turpentine	I	I
n-Propyl Alcohol	II	I	Dowclene WR	II	I
Isopropyl Alcohol	II	I	Chlorothene NU	II	I
n-Butyl Alcohol	I	I	Varsol	I	I
Isobutyl Alcohol	I	I	Inhibisol	I	I
Glycerol	I	I	Cobehn	I	I
0·9% Benzyl Alcohol	I	I	Freon TF or PCA	I	I
			Stoddard Solvent	I	I
Ether Alcohols					
Polyvinyl Alcohol	I	I			
Carbitol	IV	I			
Methyl Cellosolve	IV	I	**Cryogenics**		
Butyl Cellosolve	IV	I	Liquid Oxygen	I	I
			Liquid Nitrogen	I	I
			Liquid Helium	I	I
			Liquid Hydrogen	I	I
Ketones					
Acetone	IV	II			
Methyl Ethyl Ketone	IV	II	Glacial Acetic Acid	IV	I
Methyl Isobutyl Ketone	IV	II	10% Acetic Acid	I	I
Cyclohexanone	IV	II	6n HCl	I	I
Diacetone Alcohol	IV	I	**Acids**		
			6n H_2SO_4	I	I
			6n HNO_3	II	I
Esters			93% H_2SO_4 (Conc.)	IV	I
Methyl Acetate	IV	I	36% HCl (Conc.)	IV	I
Ethyl Acetate	IV	II	70% HNO_3 (Conc.)	IV	I
Butyl Acetate	IV	II			
Ethyl Formate	IV	II			
Methyl Formate	IV	I	**Alkalis**		
			6n NaOH	IV	I
Ethers			6n KOH	IV	I
Diethyl Ether	I	I	6n NH_4OH	I	I
Isopropyl Ether	I	I			
1, 4 Dioxane	IV	II			
Ethylene Oxide	III	I			

Test Conditions:
Total immersion of premeasured 47 mm diameter filter disc in test fluid for 72 hours at 25°C. Note: ratings may not be valid for conditions of dynamic flow through the filter or for different temperatures. This table should be used only as a general guide line. A dynamic flow test should be run if there exists any doubt regarding the applicability of static flow ratings.

Compatibility Key
For MF-Millipore:
I   Compatible—no chemical effect on filter.
II Compatible—slight swelling or distortion, but satisfactory for fluid cleaning and sterilising.
III Generally unsatisfactory—slow attack on filter.
IV Incompatible—filter dissolves or disintegrates.
* For Solvinert:
I   Compatible—with less than 15% linear swelling.
II Compatible—with 15 to 25% linear swelling.

gives details of the chemical resistance of standard MF-Millipore filters. It can be seen that low MW alcohols, ketones and certain other solvents attack this filter and thus it is not suitable for the sterilisation of alcohol-based biological materials for example. To satisfy this need, Millipore developed the Solvinert filter which is chemically resistant to virtually all known solvents and chemicals —see *Table 2.3*. The Solvinert filter is available in a 0·25 μm sterilising grade.

SELECTION OF THE FILTRATION SYSTEM

The GS (0·22 μm) is recommended for all sterile filtration applications. The GS filter is known to be capable of removing all bacteria from solutions. However, the planning of a filtration system does not end with the selection of the terminal filter type and choice of filter holder. Adequate prefiltration is often required and the optimum size and choice of depth prefilter has to be made. The next step therefore is the incorporation of the components into a satisfactory system. Two general systems, pressure and vacuum, are available and have been used. Pressure filtration offers so many significant advantages for sterile filtration, however, that it must be the recommended system. Among its advantages are the following:

1. Higher flow rates and thus faster filtration. Vacuum systems are limited to only one atmosphere of differential pressure. It has been shown that the efficiency of the membrane filter is independent of the pressure differential so to use a high pressure differential is obviously an advantage in the speeding up of the process.
2. Since the system is pressurised during filtration, any leakage is outward, reducing the likelihood of contamination. One of the serious disadvantages of the vacuum filtration system is that the sterile collecting vessel is held under vacuum to effect the filtration and thus any tiny passages between the reservoir and the outside are likely to become channels of possible airborne microbiological contamination of the filtrate.
3. No intermediate vessel is required. The liquid being processed can be filtered directly into the storage vessel or product container and the problem of dispensing the filtrate aseptically from the vacuum flask is eliminated.
4. It eliminates the possibility of contamination resulting from blow-back through the vacuum pump. This has been shown to be quite a common occurrence when using the smaller laboratory vacuum/pressure pumps.
5. Considerable foaming occurs when filtering a proteinaceous solution under vacuum, and this foaming is likely to cause denaturation of the protein. This phenomenon does not occur when using a pressure filtration system.
6. The filtration of solvents of low boiling point is made possible. In a vacuum system the low boiling solvent will be continuously evaporating at a very fast rate from the filtration reservoir.
7. A pressure filtration system simplifies the bubble point testing of the terminal filter (see next section on bubble point testing, page 32).

Some commonly used sterile filtration systems are diagrammatically represented in Figs. 2.4 and 2.5. Fig. 2.4 shows the use of a stainless steel pressure vessel which acts as a reservoir for the liquid being filtered. The vessel is pressurised by either nitrogen or air and thus forces the liquid through the filter holder for collection in a sterile receiving flask incorporating a vent filter to allow the displaced air to escape. An actual filtration system is shown in Fig. 2.6 where a bell filling system is being used to fill several small bottles with media. Fig. 2.5

shows diagrammatically a larger process system where the liquid being filtered is pumped from the batch tank through the filter holder and into a sterile receiving vessel which, once more, has a vent sterilising filter to allow displacement of air during filling and replenishment of air during emptying. Such a system is typical of that which is used by many pharmaceutical companies in the bulk sterilisation of injectable products; or by research institutes for the large scale sterile

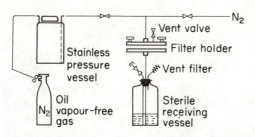

Fig. 2.4. A simple system used for the sterile filtration of small volumes of solution

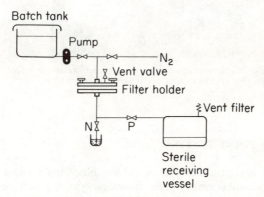

Fig. 2.5. When the receiving vessel does not readily permit viewing of the filtered solution, bubble point tests may be performed in an open beaker as shown

preparation of tissue culture media or bacteriological growth media. Neither Fig. 2.4 nor Fig. 2.5 shows a separate depth prefiltration system as this is normally only required for the sterilisation of serum or other protein solutions. In nearly all other cases of sterile filtration the depth prefilter is of such a size that it can be incorporated into the membrane filter holder by direct placement on top of the GS sterilising filter. The section on page 36 et seq. deals with a large range of liquids that can be filter sterilised and gives complete details on specific prefiltration systems.

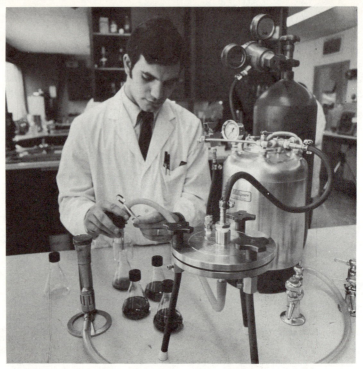

Fig. 2.6. A pressure vessel is used as a liquid reservoir before filtration. A bell filling system is used to fill several small bottles

BUBBLE POINT TESTING

A membrane filter saturated with liquid will block the passage of air or gas at pressures that would normally cause flow through a dry filter. As the pressure is increased, however, some point will finally be reached at which the surface tension forces in the larger pores will be overcome and the liquid will be expelled from these pores. If the test is performed with the outlet tube submerged in liquid, bubbles will suddenly appear at the outlet when this pressure level is reached—hence, the term 'bubble point'. *Table 2.4* lists bubble points of various Millipore membrane filters. If the solution being filtered is not aqueous, the bubble point has to be determined from tests with that solution and recorded for future use. Bubble point tests are a valuable means of checking membrane filters to be sure that there are no passages larger than the stated pore size in the filter. This test gives the assurance that after the preparation and autoclaving of the apparatus, no damage has occurred to the filter. Since it is a non-destructive test that will not endanger sterility or waste any of the product to be filtered, bubble-point tests can be performed at any time before, during or after the filtration run; but it is always wise to bubble point filter apparatus both after autoclaving and at the conclusion of the filtration run. It should be realised, however, that bubble point testing is no substitute for a sterility test of the filtrate for whilst the filter might be preventing the passage of micro-organisms it is possible that the receiving vessel was not properly sterilised or some other incidence that would cause a contamination of the filtrate. It is interesting to note that the bubble point tests cannot be usefully applied to depth filtration

Table 2.4   BUBBLE POINT PRESSURES FOR MF-MILLIPORE FILTERS WITH WATER

MF-Filter type	Pore size (μm)	Bubble point
RA	1·2	12 p.s.i.
AA	0·8	16 p.s.i.
HA	0·45	32 p.s.i.
PH	0·30	40 p.s.i.
GS	0·22	55 p.s.i.

medium for several reasons. Firstly, the average passage size in a depth filter can be as large as ten times the nominal filtration efficiency of that filter. Thus the confirmation that there are no passages larger than the average passage size in a depth filter still does not give any assurance that it will remove all bacteria from the fluid being processed. In fact, the average passage size in a depth filter is the mean of two extremes as there are always some very large passages and some very small passages within a depth filter medium.

A bubble point test can be simply carried out on all filtration systems and the procedure which would be used for the system shown in Fig. 2.4 is as follows:

1. Autoclave the filter holder, receiving bottle, inlet and connecting tubes as a single unit.
2. Connect the assembly to the pressure vessel outlet.
3. Using low pressure, allow a small volume to filter into the receiving vessel, thoroughly wetting the filter and immersing the end of the filter outlet tube. It is necessary to bleed all air from the filter holder by using the vent valve which is placed on top of all standard filter holders (Millipore).
4. Shut off flow and allow the regulated gas to pressurise the filter inlet. The regulated pressure should be equal to the operating filtration pressure.
5. Gradually increase the gas pressure and watch the outlet tube for continuous bubbling. The pressure at which this occurs is the bubble point pressure and should be recorded.
6. On completion of the bubble test, shut off the gas supply and continue filtration. At the conclusion of the filtration, all flow will abruptly stop when the gas reaches the membrane filter.
7. Carry out the bubble point test as described and observe the outlet tubing again for bubbling.

Fig. 2.5 illustrates the system using a receiver vessel that does not readily permit viewing of the outlet tube for bubble testing. With this system, use the following procedure:

1. Autoclave the assembly as a unit or sterilise the receiver vessel separately and connect it aseptically.
2. Connect the filter holder to the pressure vessel outlet. Filter a small volume of liquid to thoroughly wet the filter.
3. With clamp P closed and clamp N open, run a bubble test in a beaker of 70% ethanol.
4. After completing the bubble test, clamp off N and open clamp P to admit the process solution for filtration.
5. After the completion of the filtration, run another bubble test in the beaker.

OPERATING PROCEDURES FOR MEMBRANE FILTER HOLDERS

A complete operating procedure includes filter loading, autoclaving, connecting tubes and apparatus, pressure and bubble point testing, filtration, breakdown, cleaning and drying. Careful attention must be given not only to the steps involved in filtration, but to the preparation and handling of filters and apparatus as well. Before assembly of the filter holder it is imperative that all the components are thoroughly dry as damage can occur to the membrane filter during the autoclaving procedure if it is initially wet. Membrane filters are normally supported on photo-etched stainless support screens and these can be effectively dried by blowing them with a clean oil-free gas from a pressurised cylinder. Normally, membrane filter holders are manufactured from stainless steel which, contrary to its name, is subject to corrosion under certain conditions. Corrosion can be a severe problem if the filter holder is left uncleaned and wet for long periods of time, particularly if the wetting solution contains a high concentration of chloride ions. If corrosion should be found anywhere on the surface of the filter holder then this area should be chemically treated to remove the corrosion products and to repacify the surface at which the attack took place. Under no circumstances should abrasive substances be used to remove the corrosion products as this will only tend to further remove the passive surface of the stainless steel and therefore give rise to widespread corrosion.

FILTER LOADING

When the filter holder has been prepared for loading, the filter should be carefully placed on the photo-etched support screen and if a prefilter is also to be used this must be positioned concentrically on the membrane filter making sure that it remains completely inside the gasket sealing circle. The 'O' Rings used for sterilising filter holders are normally made from PTFE which is not a flexible substance like rubber and should be carefully checked prior to use. It is important to ensure that no small sharp fragments have been embedded in the

PTFE 'O' Ring during the cleaning cycle as these will cause a pierce hole in the filter and require reloading with a new filter. The 'O' Rings should be carefully inspected to ensure that the sealing surface has not been flattened into a wide area such that an effective seal can no longer be made. After placing the two halves of the filter holder together the unit will be sealed either by tightening locking bolts or by firmly screwing two halves together. If pressure tubing is to be used then care must be taken in its selection. Surgical latex and all-black rubber tubing are not acceptable because of the extractables they contain. Pure gum rubber or its equivalent and some polythene and PVC tubing are satisfactory and will not detrimentally affect the membrane. PTFE lined tubing and some silicone rubbers are also acceptable. All new tubing should be coiled in a pan of water and autoclaved for one cycle before use. It is naturally important to ensure that the tubing being used will withstand the pressures that will be experienced during the filtration cycle.

AUTOCLAVING

For autoclaving, small filter holders may be individually wrapped in a good quality Kraft paper or its equivalent. Larger apparatus should be autoclaved with hoses attached and hose endings wrapped as shown in Fig. 2.7. Apparatus

Fig. 2.7. 293 mm filter holder prepared for autoclaving

may be autoclaved with the receiving vessel connected; however, the vessel should not be larger than 2 litres and an adequate vent must be provided to allow the penetration of moist heat. Autoclaving should be carried out at 121°C for the period of time indicated for the particular holder being used. Autoclave temperatures are known to differ, frequently, from the indicated temperature. It is good practice to periodically place a maximum registering thermometer on the filter holder to check the accuracy of indicated temperatures. After autoclaving the apparatus is ready for use as soon as it cools to room temperature or to the temperature at which filtration will be carried out. The holder may be rapidly cooled using cold water. During autoclaving the locking bolts may slacken and it is important that they are not retightened when the apparatus is first removed from the autoclave. They should only be retightened when the filter holder is cooled and after the process solution has been admitted to the unit.

CLEANING

When the filtration run has been carried out the filter holder should be disassembled and cleaned immediately. A filter holder should not be allowed to stand overnight wet with the product. All components of the holder should be cleaned with a sponge using hot water and a liquid detergent cleanser. All threaded parts, 'O' Ring grooves, recesses and orifices should be scrubbed with a stiff bristled brush and cleanser to remove all traces of contamination. As mentioned above, never use steel wool or abrasive material on any part of the filter holder. After thorough cleaning all parts should be rinsed with hot running tap water and then rinsed several times with cold distilled or deionised water before being left to completely air dry. Wiping with cloth or paper will leave traces of fibres and lint that could contaminate the filtrate.

STERILISING FILTRATION OF LIQUIDS

Obviously it is not possible to give complete details of the methods normally used to process all types of liquid. However, it is possible to give the basic principles on the selection of a sterilising filtration system for a wide range of liquids. As mentioned earlier, all sterilising filtration systems should terminate in a GS (0·22 μm) filter which will act as the absolute sterilising medium. Obviously, the pretreatment that is necessary to enable the processing of an economic volume through a terminal GS filter will vary considerably depending on the nature of the liquid. Before deciding on the arrangement of the filters and apparatus, it is advisable to make a comparison between the liquid to be filtered and a liquid whose filtration mechanics are well known. As an aid to evaluating the filtration requirements for a particular liquid, it is possible to group all liquids into three classes depending on the pretreatment that is necessary.

(i) LIQUIDS EASY TO FILTER

The first class of liquids are those which are easy to filter and this includes nearly all clear aqueous solutions, distilled and deionised water, culture media without serum, alcohols, vitamin preparations, and many more similar liquids. All these solutions may be filtered directly through a GS filter which has a fibreglass prefilter pad resting directly on its upstream side. The fibreglass pad (Millipore Type AP25) consists of pure glass fibres and has the ability of appreciably extending the life of the GS filter. Fig. 2.8 shows diagrammatically the action of the fibreglass prefilter pad. Tap water can normally be directly filter-sterilised using a GS filter and fibreglass prefilter, however, if it should be

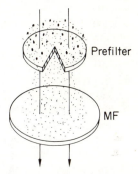

Fig. 2.8. The microfibre glass prefilters rests directly on top of the Millipore filter in the same holder

found that the water supply contains a high concentration of colloidal material, it may be necessary to regard tap water as a liquid which requires more extensive pretreatment as detailed in the next section. *Table 2.5a* gives typical filtration data for liquids that can be processed with a filtration system consisting of just a GS filter and fibreglass pad.

Table 2.5a   FILTRATION PERFORMANCE OF LIQUIDS REQUIRING A MINIMUM OF PRETREATMENT BEFORE STERILISATION

Solutions	Filter diameter mm	Filters	Performance
1. Distilled water	293	GS (0·22 μm)	4500 litres in 3 hours at 20 p.s.i.
2. Propylene glycol (80% concn.)	142	GS (0·22 μm) + AP25 Prefilter disc	14 litres in 1 hour at 30 p.s.i.
3. Hanks' Medium	293	GS (0·22 μm) + AP25 Prefilter disc	250 litres in 30 min at 15 p.s.i.
4. Petrolatum (130°F)	142	GS (0·22 μm)	20 litres in 1 hour at 80 p.s.i.
5. Sodium pentathol	142	UG (0·25 μm) + AP25 Prefilter disc	80 litres in 1 hour at 35 p.s.i.
6. Tap water	293	GS (0·22 μm) + AP25 Prefilter disc	1000 litres in 1 hour at 50 p.s.i.

(ii) LIQUIDS REQUIRING EXTENSIVE PRETREATMENT

The next class of liquids are those which require extensive pretreatment before they can be economically processed using the GS membrane filter. These liquids include those having a higher concentration of solids, haziness, or protein content, including many oils, non-synthetic bacteriological broth media, antibiotics, viral vaccines, and tissue culture media with less than 2% of unfiltered serum. These liquids generally require some form of pretreatment such as extensive prefiltration, settling or even centrifugation. Prefiltration is the most commonly used technique as centrifugation is difficult in those cases where several litres of liquid are being processed and settling is normally only efficient for larger particles which are in any case, easily removed by depth prefiltration. The prefiltration may consist of one or several grades of depth filter material such as asbestos prefilter pads, fibreglass cartridges, cotton-wound cartridges or even diatomaceous earth. Generally speaking, two types of depth prefilter are commonly used and these are the asbestos pad prefilter and cartridges consisting of either fibreglass or cotton-wound types. A typical sterilising filtration set-up consists of a fibreglass cartridge filter holder directly upstream of a 142 mm filter holder containing a GS sterilising grade filter. Such a system is used for sterilising a viral vaccine that has been allowed to settle for two hours (to separate gross cell debris) after harvesting from the growth vessel. The filtration data on this system are to be found in *Table 2.5b* which also gives

Table 2.5b   FILTRATION PERFORMANCE OF LIQUIDS REQUIRING EXTENSIVE PRETREATMENT BEFORE STERILISATION

Solutions	Filter diameter mm	Filters	Performance
1. Tissue culture media (with 2% unfiltered serum)	142	GS (0·22 μm) + AP25 Prefilter disc	20 litres in 40 min at 15 p.s.i.
2. Tissue culture media 199 (10X concn.)	142	GS (0·22 μm) + AP25 Prefilter disc	15 litres in 50 min at 12 p.s.i.
3. Clarified insulin	142	HA (0·45 μm) + AA (0·8 μm) + AP25 Prefilter disc	5 litres in 40 min at 30 p.s.i.
4. Clarified viral vaccine	293	GS (0·22 μm) + 0·5 microns, nominal fibre-glass cartridge	100 litres in 40 min at 20 p.s.i.

details of filtration data for liquids requiring extensive pretreatment. When using asbestos pad filters for prefiltration purposes, it is not necessary to rigorously adhere to the stated operating pressure differentials as it does not matter if slight breakthrough occurs, as the GS filter will prevent absolutely the passage of any bacteria. It should be noted that asbestos pads have a high adsorptive capacity due mainly to the electrostatic effect created by the inherent positive charge on asbestos fibres. This adsorptive capacity enables the asbestos pad to remove particles, micro-organisms etc. which are far smaller than the pores through the filter media. These adsorptive forces are so strong that if a viral vaccine was directly filtered through an untreated asbestos pad filter. then a very high percentage of the titre would be lost due to the adsorption of the virus particle on the asbestos fibre. To reduce this effect, asbestos pad filters may be pretreated with a gelatine solution which coats the fibres and drastically reduces the adsorptive character. However, in doing this the filtration properties are also greatly affected and the asbestos pad filter then becomes no better than a fibreglass depth filter which does not have any inherent charge and, therefore, not such a great adsorptive capacity as the untreated asbestos pad filter. Thus, in all cases where the liquid being filtered contains an 'active' component, such as a virus particle or enzyme, untreated asbestos pads should not be used as prefilters but rather a fibreglass or cotton-wound filter cartridge should be tried. Asbestos pad prefiltration is particularly useful for processing those liquids which contain small quantities of unfiltered serum. Further details on the use of asbestos pads for the filtration of serum are given in the next section which is devoted to the filtration of protein solutions, serum and tissue culture media containing up to 10% of serum. To summarise the sterilising filtration of this class of liquids, it can be said that the prefilter which is used should be carefully chosen so that it does not have a detrimental effect on the liquid being filtered.

(iii) LIQUIDS DIFFICULT TO FILTER

This section describes the filtration processes for the class of liquids which are difficult to filter. These include serum, plasma, trypsin, allergenic extracts and tissue culture media containing more than 2% of unfiltered serum. Serum is by far the most common of this class of liquids and is being processed in large quantities for many uses including that for tissue culture media.

Before giving the details of the filtration systems that are commonly used for sterilising serum, there are three general principles which should if possible, be followed when faced with the task of sterilising serum or protein solutions by filtration. These principles are:

1. That the serum or protein solution should be filtered as freshly as possible. When serum or protein solutions are refrigerated and stored for more than 24 hours, precipitation, consisting of colloidal agglomerates, will occur. These colloidal agglomerates cause the very rapid clogging of the GS ($0 \cdot 22$ μm) filter and also other larger pore sized filters. Thus if the serum can be processed before this precipitation takes place, then it will help considerably.
2. Although serum and protein solutions are normally chilled for handling and storage, they should be warmed to 30–35°C before filtration takes place. By doing this it has been found that up to four times the volume can be processed through a GS filter than when using refrigerated serum.
3. It is imperative that positive pressure is used as the filtration force so that foaming of the filtrate does not take place. It has been found that foaming which occurs during vacuum filtration of serum may cause the denaturation of proteins.
4. In those cases where the volume being processed permits, it is advantageous to centrifuge the serum or protein solution at the greatest possible force for at least 30 minutes prior to carrying out filtration.

There are two basic filtration systems which are commonly used for the filtration of serum and protein solutions; and the choice of system depends almost entirely on the volume of liquid that is being processed. For volumes less than 1 litre, a system of serial filtration is the optimum and least cumbersome technique. whilst for volumes greater than 1 litre. a pretreatment of asbestos pad filtration combined with a final membrane filter combination of $0·45$ μm (type HA) and GS ($0·22$ μm) filters gives the most economic system. In certain instances where the serum is to be used for addition to tissue culture media for the growth of particularly fastidious cells some workers have questioned the use of asbestos pad prefilters as having some effect on growth. In these cases a serial filtration can be carried out for large volumes of serum however, in most cases the economics are not so attractive as the use of asbestos pad prefilters. However, because of the doubt on the use of asbestos pad prefilters, serial filtration techniques for volumes greater than 1 litre of serum will be discussed below.

Briefly serial filtration is a technique for extending the throughput of serum, or other protein solution through a GS sterilising filter by means of prefiltering through a series of progressively finer membrane filters. The coarser grades of filters tend to remove much of the colloidal material which would rapidly clog the GS filter if it was used alone to process the serum. It has been found advantageous to use Dacron mesh separators between each pair of membrane filters as they tend to increase the usable area of each membrane and thereby extend the filtration life. The separators improve cross flow by preventing occlusion of pores in one filter by those in the other and also the contamination being removed is more evenly distributed. The separators are normally washable and reusable.

FILTRATION—VOLUMES LESS THAN 50 ML

When the volume of serum to be processed is less than 50 ml a convenient filtration system to use is a filter holder attached to the outlet of a syringe. Filter holders are available (Millipore) that can be loaded with a serial arrangement of filters of diameter size 25 mm. The filter holder has a female Luer Lok fitting for direct attachment to a male Luer Lok syringe. Fig. 2.9 shows the Millipore microsyringe filter holder being used for sterilising liquids.

FILTRATION—VOLUMES UP TO 500 ML

A convenient filtration system for the sterilisation of serum and protein solutions in volumes up to 500 ml is the Millipore stainless pressure filter holder shown in Fig. 2.10. This filter holder may be loaded with a serial arrangement of filters and then autoclaved ready to carry out the sterilising filtration. This filter holder has a barrel reservoir of 100 ml and if the volume being processed is 100 or 200 ml then it is quite convenient to fill the barrel once or twice. However, if more than 200 ml of liquid is being processed, then it is more convenient to use a pressure vessel as the liquid reservoir and force the liquid from the pressure

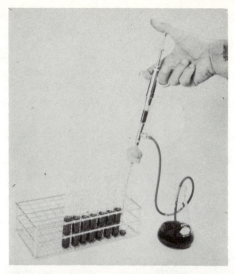

Fig. 2.9. 25 mm filter holder and continuous pipetting device being used to fill tubes with sterile medium

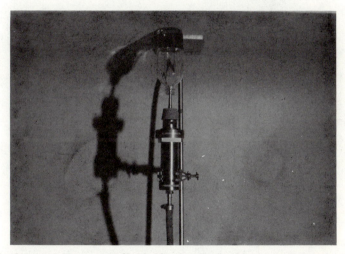

Fig. 2.10. 47 mm stainless pressure filter holder showing the 100 ml capacity barrel acting as a solution reservoir

vessel through the filter holder with pressurised nitrogen from a cylinder. A typical system for this operation is shown in Fig. 2.6.

FILTRATION—VOLUMES BETWEEN 500 ML AND 5 LITRES

For volumes greater than 500 ml but less than 5 litres, either the 90 mm sterilising filter holder or the 142 mm sterilising filter holder may be used in conjunction with a pressure vessel acting as the liquid reservoir. Such a system is shown in Fig. 2.11 where a 142 mm sterilising filter holder (Millipore) is seen being used in conjunction with a 4 litre pressure vessel.

Recently Millipore introduced liquid reservoirs for their 90 mm and 142 mm sterilising filter holders by placing a tube of plastic or stainless steel between the top and bottom plates of the filter holders. The 142 mm filter holder with liquid reservoir is shown in Fig. 2.12. The volume of these reservoirs is up to 1·5 litres

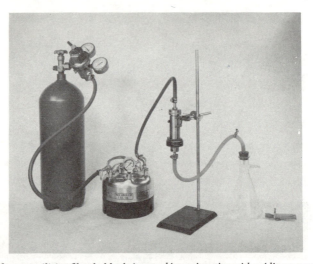

Fig. 2.11. 142 mm sterilising filter holder being used in conjunction with a 4 litre pressure vessel

Fig. 2.12. 142 mm filter holder with 2 litre reservoir for liquid to be filtered

and therefore they provide a convenient system for sterilising a few litres of serum without the use of a pressure vessel. A serial filter arrangement may be incorporated and the whole unit autoclaved prior to carrying out the sterilising filtration.

FILTRATION—VOLUMES UP TO 100 LITRES

When much larger volumes of serum are being processed, say up to 100 litres, it has been found that an asbestos pad filter will reduce the need for the coarser grades of membrane filter in the serial set-up and only the $0.45\,\mu m$ and $0.22\,\mu m$ filters are required. Normally it is not convenient to use the asbestos pad filters for much smaller volumes due to the lack of suitable laboratory scale asbestos pad filters capable of processing approximately 1 litre of solution. The use of asbestos pads for the filtration of serum has some advantages over other depth filter media due to the attraction of the asbestos fibre for the colloidal protein agglomerates which so rapidly clog the GS membrane filter. A typical filtration set-up to cope with 25 litres of serum would be the use of a filter press containing approximately five 14 cm pads of fine clarifying grade asbestos pad

Table 2.5c FILTRATION PERFORMANCE OF SOLUTIONS CONTAINING HIGH CONCENTRATIONS OF PROTEINS

Solutions	Filter diameter mm	Filters	Performance
1. Calf serum (asbestos pad prefiltered)	293	GS ($0.22\mu m$) + HA ($0.45\mu m$)	30 litres in 2 hours at 40 p.s.i.
2. Clarified plasma (asbestos pad pre-filtered)	142	GS ($0.22\mu m$) + HA ($0.45\mu m$)	6 litres in 90 min at 30 p.s.i.
3. 4% Gamma globulin	142	GS ($0.22\mu m$) + HA ($0.45\,\mu m$) + AA ($0.8\,\mu m$) + SM ($5.0\,\mu m$) + AP25 Prefilter disc	10 litres in 1 hour at 20 p.s.i.
4. 25% Gamma globulin	293	GS ($0.22\,\mu m$) + HA ($0.45\,\mu m$) (asbestos pad prefiltered)	53 litres in 2 hours at 20 p.s.i.

(such as type S 10 Carlson-Ford) used as a pretreatment for a $0.45\,\mu m$ and $0.22\,\mu m$ (GS) filter set placed in a 293 mm diameter membrane filter holder. With this system, approximately 30 litres of serum may be processed in 2 hours using a 45 p.s.i. pressure differential across the complete system. Fine clarifying grades of fibreglass prefilters or even diatomaceous earth filter media do not have nearly as great prefiltration efficiency as the asbestos pads. *Table 2.5c* gives details of the filtration data for other protein solutions.

To carry out the sterilising filtration of tissue culture media containing up to 10% of serum, it has been found that the optimum processing technique is as follows. The media should first be sterilised with a filtration system consisting simply of a fibreglass prefilter pad placed on top of the GS sterilising membrane filter. The serum should also be processed separately using the techniques which have been described above. After the two process solutions have been mixed then the final sterilising filtration can be carried out using a filter holder containing the fibreglass prefilter pad placed on top of the GS sterilising membrane filter. If the serum is held for more than 48 hours before being added to the media and the whole being processed, it is advisable to use a fine clarifying grade of asbestos pad prefilter before the GS membrane filter to remove the colloidal protein agglomerates which normally form in serum on standing for lengthy periods. An example of the filtration data on the latter system is that 200 litres of tissue culture medium containing 10% of serum may be processed in one hour using a system consisting of a 293 mm diameter asbestos pad (type S10 Carlson-Ford) and a single 293 mm diameter GS membrane filter. A system pressure of 30 p.s.i. was employed.

SIZING THE FILTER HOLDER

The efficient planning of a filtration system cannot be carried out with an unknown liquid until some filtration data have been collected. One simple way to do this is to take a small filter holder, say the 25 mm microsyringe filter holder, and determine how much of the solution to be processed can be passed through this relatively small area at a pressure that can be obtained on the large scale arrangement. The filtration area required for the large batch may then be calculated as follows:

$$\frac{A}{a_s} = 1 \cdot 5 \, \frac{V+}{v_s}$$

or

$$A = 1 \cdot 5 \, \frac{V+}{v_s} \, a_s$$

where: v_s = sample volume (ml)
  a_s = area of sample holder (cm^2)
  $V+$ = production batch volume
  A = required area for production sterile filtration
  $1 \cdot 5$ = safety factor

The filtration areas of available membrane filter holders may be found in *Table 2.6*. Fig. 2.13 shows a system employing a polypropylene 25 mm diameter filter holder (Swinnex-25) being used to calculate the correct area of filter for large process systems.

Table 2.6 FILTRATION AREA FOR VARIOUS DIAMETER FILTER HOLDERS

Filter holder	Filter diameter (mm)	Prefilter diameter (mm)	Area Eff. (cm²)	Autoclave time, min.	Conns.
SW-13	13	10	0·7	12–15	Luer
SW-25	25	22	3·4	12–15	Luer
SW-47	47	35	13·8	15	Luer & tube
Swinny	13	10	0·8	15	Luer
Micro-syringe	25	22	3·9	15	Luer
Sterifil	47	35	13·8	20	$\frac{1}{4}''$ hose
47 mm	47	42	11·3	20–25	$\frac{7}{16}''$ hose
90 mm	90	75	39	20–25	$\frac{7}{16}''$ hose
142 mm	142	124	97	25–30	$\frac{7}{16}''$ hose
293 mm	293	257	486	35–45	$1\frac{1}{2}''$ T.C. or $\frac{9}{16}''$ hose

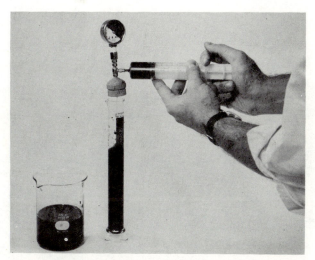

Fig. 2.13. The size of a filter holder needed to process a given volume of solution can be quickly calculated using a filter sizing apparatus

GAS STERILISATION

In gas filtration, the enormous specific surface and high resistivity of membrane filters creates substantial electrostatic charges that prevent the passage of particles far smaller than the pore dimensions. As a result, the efficiency of membrane filtration is increased to such a degree that a 0·5 μm filter will ensure the sterility of gases. Millipore has developed a hydrophobic version of its Solvinert filter that offers significant advantages for the filtration of air and other gases. This filter, whose chemical structure has been treated so that it will resist wetting by water, will not become saturated and airlock. Water intrusion pressures for the hydrophobic Solvinert filter closely correspond to the Bubble Point Pressure of the standard filter. The hydrophobic Solvinert filters may be autoclaved freely or sterilised in place with live steam and will resist dry heat to 200°C.

Commonly used depth filters, such as fibreglass, packed cotton, sinters etc., suffer from many disadvantages when being used for gas sterilisation. The random orientation of the filter media often results in channelling, or passages of relatively easy flow. These may result in wholly inadequate filtration, especially during violent vacuum release such as occurs when a vacuum pump is disconnected. Grow-through can occur when the medium becomes moistened by filtrate carry-over or vapour that turns it into an excellent nutrient bed.

During filtration, fluids are transferred from one vessel to another, requiring that both vessels be vented to the atmosphere. During sterile filtration, the receiver vessel should always be vented through a sterilising filter to block the entry of airborne contamination. Membrane filters loaded into suitable filter

Fig. 2.14. Swinnex-47 filter holder being used as a sterile vent on a receiver vessel

holders are ideal for both of these applications. The Swinnex-47 filter holder (Millipore) is especially convenient for the small scale and laboratory systems and it has the added advantage that the filter is supported in both directions so that vessels can be emptied and filled with complete security. Fig. 2.14 shows a Swinnex-47 filter holder with a hydrophobic Solvinert filter being used to vent a small pressure vessel.

SUMMARY

Sterilisation filtration has been made a much more predictable process with the introduction of membrane filters due to their high uniformity and repeatability.

It is feasible to economically sterilise virtually any fluid with a membrane filter providing that the suitable pretreatment process has been selected.

REFERENCES

1. FLOOD, J. E., PORTER, H. F. and RENNIE, F. W., 'Filtration Practice Today', *Chem. Engng, Albany,* **73**, 163 (1966)
2. OSGOOD, G., 'Filter Sheets and Sheet Filtration'. Presented to the Filtration Society, London, September 1966
3. GARVAN, J. M. and GUNNER, B. W., 'The Harmful Effects of Particles in Intravenous Solutions'. *Med. J. Aust.,* **2**, 1 (1964)
4. PORTNER, D. M., PHILLIPS, C. R. and HOFFMANN, R. K., 'Certification of probability of sterilisation of liquid by filtration', *Appl. Microbiol.* **15**(4), 800 (1967)

THREE

Techniques for cell cultivation in plastic microtitration plates and their application in biological assays

MAX J. ROSENBAUM, ELIZABETH J. SULLIVAN and EARL A. EDWARDS
Naval Medical Research Unit No. 49, Great Lakes, Illinois, U.S.A.

INTRODUCTION

Conventional methods for the cultivation of cells usually employ glass vessels. The glass screw-capped tube is the most common culture unit used for virus assay or serological tests which are carried out in cell cultures. Such cultures require substantial volumes of expensive media and cells, much of which is unnecessary, for the proper function of the test. These tubes are cumbersome to handle and require large areas of bench and incubator space. Another disadvantage is that removal and replacement of caps is not only tedious, but can be dangerous due to glass breakage. Therefore, such operations are time consuming, uneconomical and require a large staff of technicians, and few virus laboratories are able to support wide-scale epidemiological studies.

Some of these disadvantages have been overcome by using cells grown in plastic panels. Barski and Lepine[1] devised an acrylic plate which supported the growth of cell monolayers. These cells could be infected with poliovirus and the cytopathology could be observed microscopically. Serum antibody neutralisation tests, as well as infectivity titrations, could be carried out in these plates. Later, Melnick and Opton[2] reported that opaque plastic trays with moulded depressions or cups could be employed for colorimetric assays of poliovirus antibody. Johnston and Grayston[3], and Lennette et al.[4] devised a colorimetric test for adenovirus neutralising antibody using plastic trays. Rightsel et al.[5] have used a panel composed of a soft polyvinyl plastic containing cups for a variety of viral tests and suggested its use for screening of anti-viral drugs. Since this material was transparent, microscopic methods could be used to observe the results of these tests.

The advantage of utilising plastic panels are obvious. They are less expensive and their manipulation for cell cultures and viral tests is easier, faster and more economical than for conventional tube tests. They require less space, can be discarded after use, thus eliminating the need for elaborate decontamination and cleansing facilities.

An innovation which enhanced the advantages of plastic containers was the adaptation of the Takatsy microtitration technique[6] for tissue culture purposes. Originally designed for viral serological tests, which did not employ tissue culture cells, the technique was modified by Sever[7] for the colorimetric assay of poliovirus in cell suspensions. Rosenbaum and co-workers[8, 9] demonstrated that monolayers of various cell types could be established in either hard (Plexiglas) or soft (polyvinyl) microplates (Cooke Engineering Co., Alexandria, Va.). The microplate cultures have proved useful for a variety of virological and cytological studies and produced results comparable to conventional macro methods.

The principle of the Takatsy microtechnique is ingeniously simple. Most assays of biological fluids involve making increasing dilutions of a constant volume of active material (dilutant) in a constant volume of diluent. The two-fold dilution (equal parts of dilutant and diluent) is most commonly employed for serum assays, and the tenfold dilution is often (but not exclusively) used for virus titration. Whereas conventional dilutions utilise the glass pipette to deliver, mix and retrieve the dilutant, the micro system employs a 'loop' which performs a similar function with as little as 0·025 ml of dilutant. This volume is added to an equal or greater volume of diluent provided by a micropipette. Thus, dilutions can be accomplished in far less time than is required by macro methods and require only a fraction of the volume of reagents.

These advantages, coupled with a great capacity to easily produce replicate micro cultures of cells and conserve all constituents and efforts involved, make this system extremely useful for many situations where cell cultures are utilised.

The following section will discuss the present knowledge of the preparation of tissue cultures in microplates, how they are utilised, and suggest potential areas in which they may be adapted.

MICROTITRATION MATERIALS AND TECHNIQUES

A. EQUIPMENT

The basic implements for the microtitre system for tissue culture and assay procedures are shown in Fig. 3.1. These include the microplate, micropipette (dropper), diluting loop (diluter) and priming blotter.

The plates, droppers and diluters are available from several commercial sources (Cooke Engineering Co., Alexandria, Va.; Microbiological Associates, Bethesda, Md.; and Linbro, New Haven, Conn.). Recently, a third supplier (Baltimore Biological Laboratories, Cockeysville, Md.) is marketing, not only a modified plate, but will also provide them with established monolayers of various cell types.

1. PLATES

Plates are available in several types of plastic. The original type, composed of rigid acrylic (Plexiglas) is relatively expensive, but is reusable. These are commonly used for complement fixation (CF) or haemagglutination (HA) tests, but can be employed for tissue cultures. An alternate type of plate is made of a pliable polyvinyl plastic. It is inexpensive and can be discarded after use. Most of the techniques which will be described pertain to this type of plate. The plates can be obtained in various sizes and with different numbers and arrangements of wells. The bottom of the well is the surface upon which the cell monolayer is established. Surface contours of the bottom are available in either a 'U' or flat form. The former is most commonly used in tissue culture procedures which do not require high resolution microscopy (over 100×). The latter, because of its planar qualities, is desirable for this type of work. The plate shown in Fig. 3.1 is 8 × 12 cm and contains 96 'U'-type wells in rows of 8 × 12.

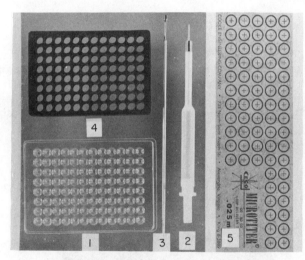

Fig. 3.1. Equipment for microtitration procedures.
1. Microplate for cell cultures
2. Micropipette (dropper); delivery—0·025 ml
3. Transfer loop (diluter); delivery—0·025 ml
4. Rubber plate holder
5. Priming blotter

Each well has a capacity of approximately 0·25 ml. A rubber plate holder (Fig. 3.1) is used to provide support for the pliable plate when applying or removing taped coverings.

Often when plates arrive from the manufacturer they are toxic for cells. This toxicity can be removed by soaking the plates in 70–90% ethyl or isopropyl alcohol for 1–2 hours. (Some investigators use an additional preliminary detergent wash, but we have not found this necessary). The alcohol should not be allowed to evaporate since this fixes the toxic property. After the alcohol bath, the plates are inverted to drain and are rinsed in three changes of tap water, followed by three rinses in triple-distilled water. They are then drained thoroughly, air dried and stored in their original boxes until used.

Plates are not routinely pre-sterilised since contamination does not appear to be a problem when tissue culture media contain antibiotics. Where such problems exist or where absolute sterility is essential to the test, preliminary plate sterilisation can be accomplished by either ultraviolet irradiation or ethylene oxide gas.

Wells can be covered by applying a transparent adhesive tape (Paklon, 3M Company, Minneapolis, Minn.) to the top surface of the plate or by overlaying 1–2 drops of light-weight mineral oil (Drakeol, Pennsylvania Refining Co., Butler, Pa.) to the media in each well. When cell cultures are cultivated under CO_2 atmosphere, in a well-humidified incubator, light-weight non-adherent covers will suffice to keep out dust.

2. MICROPIPETTES

These pipettes or droppers (Fig. 3.1) are moulded from a rigid plastic (polypropylene) and will withstand autoclave temperatures. They are tipped with needles with an orifice size that delivers either the standard drop of 0·025 ml or a larger drop of 0·05 ml. These droppers are used to deliver any fluid material which is added to the wells in a constant amount, such as diluent, cell

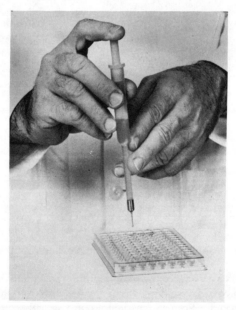

Fig. 3.2. Dropping technique for micropipette. (Reprinted by permission of Microbiological Associates, Bethesda, Maryland)

suspension, or virus dosage. The dropper can be fitted with cotton mouthpiece plugs, as a safety precaution. Some investigators have found that inexpensive Pasteur pipettes with a calibrated orifice will also suffice as droppers and can be discarded after use. In use, the fluid is drawn up into the dropper by suction and the needle tip is wiped with sterile tissue or gauze to remove excess fluid insuring even sized drops. The rate of drops is usually controlled by finger manipulation. A rubber bulb may also be used for this purpose (Fig. 3.2). Droppers should be held in a vertical position to insure accurate delivery.

3. LOOP DILUTERS

The unique innovation of the microtitre system is the loop, which is used to dilute and transfer the material to be assayed. Essentially, the loop is a modification of the bacteriological loop which is used to transfer an exact volume of fluid. When the loop is dipped into a fluid and removed, a film adheres across the ring due to capillary and surface tension forces. Such forces are

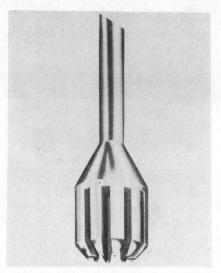

Fig. 3.3. Tulip-type diluter—stainless steel (5 × actual size). (Reprinted by permission of Microbiological Associates, Bethesda, Maryland)

dissipated when the loop is reintroduced into a volume of liquid and the film is released into that liquid, and a dilution effect is made. Successive transfers of the loop into fresh fluid volumes of constant amount with a constant carry-over of dilutant accomplishes the dilution titration.

Several versions of the loop have been constructed. The original Takatsy loop was a continuous length of coiled wire rounded into a ball form. The most recent loop design is the so-called tulip-type (Fig. 3.3) and is an improvement over the coil, since it is less apt to sustain damage in use. Either type will transfer 0·025 ml volumes (0·05 ml diluters are also available). The diluters are stainless steel and can be flamed for cleansing and sterilisation.

Dry diluters will not adsorb a correct amount of fluid. Prior to use, they must be primed with sterile saline or other diluent before they will accurately transfer the material which is being assayed. The blotter with imprinted rings (Fig. 3.4)

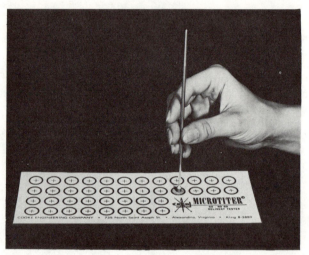

Fig. 3.4. Blotter test for adequate diluter priming. (Reprinted by permission of Microbiological Associates, Bethesda, Maryland)

is used to check for adequate priming. Saline wetted loops are touched to the centre of an unused ring circumscribed on the blotter. The fluid is absorbed by the blotter and a darkened area will flow to the circumference of the ring (Fig. 3.4). The residual film remaining on the loop acts as a primer when the loop is introduced into the fluid being assayed. This procedure, however, does not assure that diluters have been accurately calibrated.

B. DILUTER CALIBRATION

A simple procedure for diluter calibration and adjustment of diluters by refractive index methods has been described by Edwards[10]. Calibration is accomplished by making dilutions of serum in saline and determining the refractive index of each dilution. Several types of inexpensive refractive index instruments are available (TS Meter, American Optical Co., Buffalo, N.Y.). When the instrument readings are plotted on logarithmic paper, a straight regression line indicates that the diluters are accurately adjusted. If not, corrections in the loop opening should be made with forceps or another tool.

Calibration checks should be performed on new diluters and periodically after service.

C. DILUTION CYCLE

To enable the operator to maintain a uniform method for manipulating the diluters, a sequence of steps (dilution cycle) was described by Sever[7]. This cycle has been modified by our laboratory and is shown in Fig. 3.5. It should be emphasised that a delay between steps 4 and 5 may result in the diluter losing its priming and produce erroneous titres. Merely touching the diluter to the surface

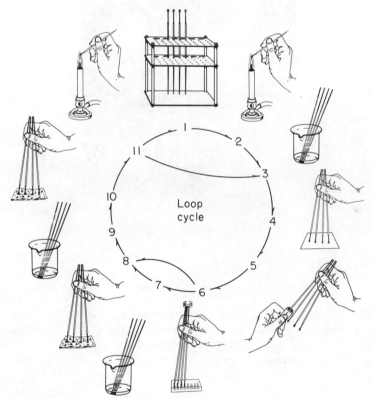

Fig. 3.5. Use of diluters for transfer and dilution of assay material (loop cycle).
1. Diluter stand.
2. Flame cleansing of diluter (allow to cool).
3. Diluter priming in diluent (1 min).
4. Blotter test for adequate diluter priming.
5. Filling of diluter with assay material.
6. Dilution of assay material in diluent by diluter rotation.
7. Diluter decontamination (viral assays) in 0·5% sodium hypochlorite solution (1 min).
8. Blot on gauze to discharge fluid.
9. Diluter rinse in distilled water (1 min).
10. Blot to discharge fluid.
11. Flame cleansing of diluter (allow to cool)

of reagent is sufficient contact for filling. Contact with bubbles should be avoided. If non-infectious material is being titrated, step 7 (decontamination) can be by-passed. Handling up to eight diluters at one time requires practice (Fig. 3.6), especially in transferring from well to well. Usually, the novice operator begins by handling four diluters. With practice, this can be increased to eight. Thus, an entire plate of eight samples can be titrated simultaneously.

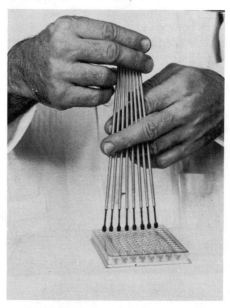

Fig. 3.6. Dilution technique using eight diluters simultaneously. (Reproduced by permission of Microbiological Associates, Bethesda, Maryland)

By following the dilution cycle (Fig. 3.5), a uniform procedure is used by each operator. This is important in overall reproducibility of titre determination. Diluters can be used, not only to produce constant decrements of virus or serum, but can be used to dilute cell suspensions, assay toxins, antitoxins, anti-viral drugs or any other desired material in suspension or solution form.

D. MICROSCOPES

Results of the various assays can be observed by either micro or macro methods. A microscopic examination is generally the most sensitive method to evaluate results with viruses that produce cytopathology or cytolysis. Such observations of the condition of viable cells and progress of viral cellular degeneration in microplates is made with either a conventional or inverted microscope.

The advantage of the latter instrument (Fig. 3.7) is that cells or fluids will not be disturbed during observation. This is mandatory when sealing tape is not employed. In other situations where the plate may be inverted without damage to the contents, an ordinary microscope can be used.

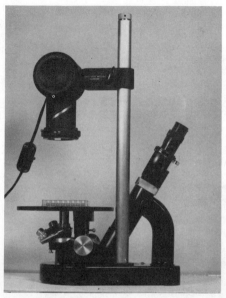

Fig. 3.7. Inverted microscope with microplate on stage. (From Sullivan and Rosenbaum[9], courtesy of the publisher of Am. J. Epidem.)

Titre endpoints of virus tests are often determined colorimetrically by the metabolic-inhibition technique. Changes in the colour of a pH indicator (non-toxic for cells), which is incorporated in the culture media, are used to evaluate virus growth or inhibition.

Other viral indicators, such as haemagglutination, can be used for macroscopic evaluation of tests made with myxoviruses or other agents possessing this property.

Only the basic equipment required for most tissue culture and assay procedures has been mentioned above. An automatic titration system is available commercially (Autotiter, Astec, Inc., Orange, Conn.) and may be advantageous if serological testing is routinely done on a massive scale.

PROCEDURES FOR CELL CULTIVATION

Many types of cells which grow in glass tubes can be cultivated in microplates[9]. These include primary or secondary cells from organs, semicontinuous strains, such as diploid (human embryonic lung) or serially propagated cells of malignant origin. Microsystem cell cultivation techniques are similar to macro methods, but require only a fraction of component volumes. The following procedures are only recommendations and may be modified to fit a particular need.

A. MEDIA

1. GROWTH MEDIA

The cell growth media can be similar to those commonly used to propagate a particular type of cell in glass bottles. We have found that a 'universal' growth media is satisfactory for any cell of the three basic types. The composition of this media is given below:

Composition of Nutrient Medium (100 ml)
Eagle's Basal Media[11] (BME) in Earle's Balanced
    Salt Solution[12] (EBSS)   80·0 ml
Tryptose phosphate broth   4·0 ml
Foetal or agamma calf serum   10·0 ml
$NaHCO_3$(7·5%)   1·8 ml
Penicillin and streptomycin (200 units or mcg/ml,
    respectively) mixture   0·2 ml
Neomycin (100 mcg/ml)   0·1 ml
Amphotericin-B (5 mcg/ml)   0·05 ml
EBME   Q.S. to 100 ml
pH 7·0–7·2

Calf serum is essential for the adherence and flattening of cells to the plastic surface. The minimum concentration required will vary with a particular cell line. Sera should be pretested to ensure that specific or non-specific viral inhibitors are not present in the concentration used. One serum supply source (Hyland Laboratories, Los Angeles, Calif.) will provide such inhibitor-free serum upon request.

2. MAINTENANCE MEDIA

After cells are established, they can be kept for various periods of time by changing the growth media to a maintenance type. Media changes are easily accomplished by dumping the fluid contents out with a quick, vigorous hand movement. The cells can then be washed gently with a balanced salt solution and the maintenance milieu added.

Maintenance media for assays in established monolayer can be of the chemically defined type, media 199[13] or Eagle's Basal Media[11], for primary or secondary cells. For other cell types, the universal media with a reduced serum concentration (2–5%) is satisfactory. Also, a slightly higher pH (7·3–7·4) is recommended for maintenance of cells.

An alternate maintenance media, L-15 (Liebovitz[14]), has been used in microplates by several investigators[15, 16]. In this media, cells can be kept without being sealed, and in free exchange with atmospheric gaseous conditions. This is due to the substitution of galactose for glucose as the energy source and cells metabolise at a slower rate with less CO_2 evolved.

3. DILUENT

Media for dilution of serum or virus can be any of several varieties conventionally used. However, since it comprises 30–50% of the total volume of growth media, it should be compatible with those constituents that nurture the cells. A standard diluent used by us is prepared as follows:

Composition of Diluent (100 ml)

Lactalbumin-hydrolysate (0·5%) in Earle's Balanced Salt Solution (LHE)	95·5 ml
$NaHCO_3$ (7·5%)	1·0 ml
Penicillin and streptomycin (200 units/ml, 200 mcg/ml, respectively)	0·2 ml
Neomycin (100 µg/ml)	0·1 ml
Amphotericin-B (5 µg/ml)	0·05 ml
LHE	Q.S. to 100 ml
pH 7·0	

B. CELL PREPARATION

Procedures for the cultivation of microplate tissue culture are shown in Fig. 3.8.

Suspensions of seed cells are obtained from trypsinised organs[17] or stock cultures grown in bottles. Diploid cells are dispersed by a residual trypsin technique[18]. Other methods for cell dispersal such as versene may be employed. In some cases, a vigorous shaking of the bottle will suffice.

Seed cells may also be obtained from suspension ('spinner') cultures. These are quite suitable since they are usually in a linear growth phase and form a monolayer within 24 hours after implantation.

Cell concentrations for seeding microplates are about the same as used in tube type tissue cultures, but the volume required is only one-fortieth of the latter. The cell concentrations and media recommended for the various types of cells are shown in *Table 3.1.*

Cell suspensions in growth media are added to the microplate by dropper in 0·025 ml volumes. If used for assay purposes, these cells are added after dilutions of the material being tested have been made. Since cells tend to aggregate, they should be dispersed by tapping the plate. Proper cell dispersal is checked by microscopic observation. The wells are then sealed either with two drops of oil or covered with the tape mentioned previously. This is done to

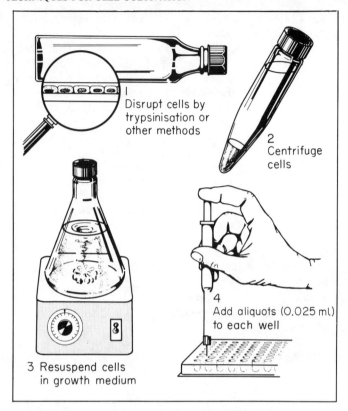

1
Disrupt cells by
trypsinisation or
other methods

2
Centrifuge
cells

4
Add aliquots (0.025 ml)
to each well

3 Resuspend cells
in growth medium

Fig. 3.8. Preparation of seed cells for microplate tissue cultures

Table 3.1. REQUIREMENTS FOR CELL CULTIVATION IN MICROPLATES

Cell type	Seed concentration	Growth media (pH 7·0)	Days until monolayer formed	Maintenance media (pH 7·3)
Primary cells (kidney, amnion, lung, etc.)	300 000/ml	'Universal'* or media of choice	5–7	EBME or 199†
Secondary strains (lung, kidney, etc.)	2:1 split		2–3	EBME +2% serum (foetal calf)
Continuous lines (HeLa, HEp-2, KB, etc.)	100–200 000/ml		2–3	Ginsberg's (1955)

* See cell cultivation.
† Eagle's Basal Media (in Earle's Salt Solution)—(Eagle[11], Earle[12]). Media 199 (Morgan *et al.*,[13]).

prevent excessive dehydration and loss of gases (CO_2) to the atmosphere during incubation in an ordinary bacteriological incubator. If a CO_2 incubator is available, a lid-type cover is all that is required to keep out dust. Increased CO_2 tension (5%) is especially desirable for diploid cell cultivation. In either type of incubator, a high relative humidity is necessary.

A useful device for obtaining atmospheres of various combinations and concentrations of gases is the Case jar (Scientific Supply Co., Chicago, Ill.) shown in Fig. 3.9. Originally designed as an anaerobic jar for bacterial culture,

Fig. 3.9. Case anaerobic jar for microplate cell cultivation in extraordinary atmospheres (1/9 actual size). (From Sullivan and Rosenbaum[9], courtesy of the publisher of Am. J. Epidem.)

it lends itself admirably for microplate cultures. The two-valve system in the cap permits replacement of gas (from tank source) after plates have been introduced into the jar. The entire system can then be placed in an ordinary incubator. Thus, a whole series of environmental conditions can be arranged at a very reasonable cost. Such equipment has been used to simulate a Martian gaseous environment and study its effect on cell growth[29].

The length of incubation will vary with the cell. Generally 2–3 days will be sufficient time for monolayer formation for all but the primary cells. Seven days is required for cells such as primary monkey kidney, but this time may be reduced by using second generation cells. Usually, microplate cell cultures can be maintained for 7–14 days provided periodic replacement of media is made.

C. EVALUATION OF CULTURES

The appearance of various types of cells after monolayer formation is shown in Fig 3.10. These photographs are low power magnifications of viable cells taken with the aid of an inverted microscope. At this magnification, the entire disc of cells which comprise the monolayer can be seen in a single field. Thus, a rapid overall evaluation can be made of the condition of each culture.

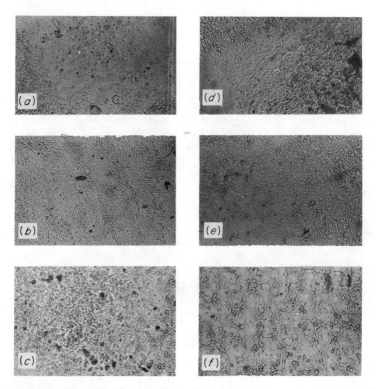

Fig. 3.10. Appearance of various unstained cell monolayer cultures in microplates (17×).
(a) Human epithelioid (HeLa)
(b) Human diploid lung (WI-38)
(c) Human epithelioid (KB)
(d) Human epithelioid (HEp-2)
(e) Monkey kidney (secondary)
(f) Rabit kidney
(RK-13). (From Sullivan and Rosenbaum[9], *courtesy of the publisher of* Am. J. Epidem.)

The appearance of the cells is quite similar to those grown in tubes. However, the formation of clumps is indicative of improper dispersal of cells.

Toxicity from either the plate itself or from any of the components of the system also produce a clumping effect with rounding and pyknosis of the individual cells.

UTILISATION OF MICROPLATE CELL CULTURES

The most common use of the microplate tissue culture has been in connection with either viral assay, typing or immunological tests of viral serum antibody. Many of these procedures have been used in epidemiological studies of polioviruses[7, 8, 19, 20], reoviruses[15], coxsackie viruses[21], rhinoviruses[22, 23], parainfluenza and influenza[24-26], rubella[27, 28], vaccinia[24] and adenoviruses[30, 31]. Several of these investigators compared the results of the micro system with macro methods. In all instances, favourable agreement was obtained and all investigators indicated the great advantages of the micro system when mass serology was involved.

A. VIROLOGICAL PROCEDURES

Many viral tests can be carried out in microplates by adding seed cells (cell suspensions from a prior monolayer) immediately after virus has been diluted, or after serum-virus mixtures have been preincubated. This is unlike other systems where the assay material is added to preformed cell monolayers. Often, there is no advantage to the established monolayer technique which requires more time, nutrient replacement and washing before the test can be initiated. Furthermore, full advantage of the microtitre system is not utilised since the twirling action of the diluters disrupts the established cell monolayer and dilutions must be carried out in an empty plate and then transferred to the one with cell cultures. Where established cell monolayers are necessary, however, microplate techniques are still much more economical than conventional tube cultures.

A typical protocol for conducting viral infectivity titrations or serum antibody neutralisation tests is shown in *Table 3.2*. The volumes of diluent shown are for twofold dilution increments, however, greater volumes can be substituted provided the total dilution volume contains sufficient nutrient for cell growth.

The virus or serum assay technique is diagrammed in Fig. 3.11a and a typical neutralisation test is shown in Fig. 3.11b.

One drop (0·025 ml) of diluent is added by dropper to a series of wells. The number of wells employed depends upon the expected titre of material being assayed. The loop is then used to pick up and insert the assay material (virus or serum) to the first well. Mixing is made by rotating the stylus five or more times. The diluted material (0·025 ml) is retrieved from the first well with the diluter and added to the next well. Dilutions are continued as desired.

In neutralisation tests, one drop of predetermined virus dose is added by dropper to each serum dilution and mixed by gently tapping the plate. The serum-virus mixtures are incubated for an appropriate period and one drop of cell suspension is added to each well and cells are dispersed by tapping. Appropriate controls such as cell, virus back-titration and serum toxicity should be included in the test. Plates are then sealed with oil or tape and incubated for either a predetermined period or until virus controls indicate the desired dosage has been attained.

Table 3.2 PROTOCOLS FOR VIRUS ASSAY OR NEUTRALISATION TESTS IN MICROPLATE TISSUE CULTURES

VIRUS INFECTIVITY ASSAY

Reagent sequence	Diluent	Virus	Procedure	Cell suspension (in growth media)	Additional diluent (optional)	Incubation procedures — Normal atmosphere	CO_2 atmosphere
Volumes (ml)	0·025	0·025	Microtitrate	0·025	0·025	Oil seal (0·1 ml) or Tape	Lid cover

NEUTRALISATION TESTS

Reagent sequence	Diluent	Serum or inhibitor	Procedure	Virus dose* (10–100 $MTCD_{50}$)	Procedure	Cell suspension (in growth media)	Incubation procedures
Volumes (ml)	0·025	0·025	Microtitrate	0·025	Mix reagents; cover plate(s) with lid; incubate (30–120 min)	0·025	As above

N.B. Appropriate controls should be included, i.e. cell (both tests) serum and virus back titration (neut. test).
Tests are incubated in ordinary or CO_2 incubator 3–7 days.
Record results by degree of cytopathology or metabolic-inhibition.
* $MTCD_{50}$ = Dosage previously calculated in microplates.

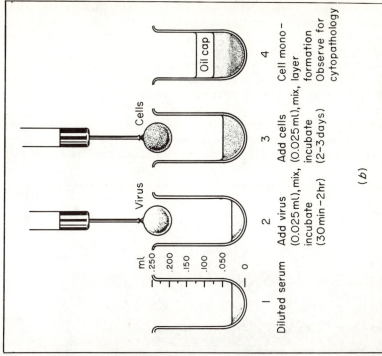

(b)

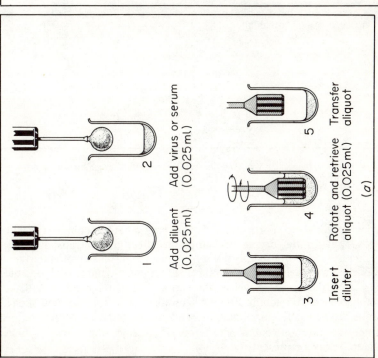

(a)

Fig. 3.11. (a) Procedure for assay of virus or serum by microtitration technique (two-fold dilution) (b) Procedure for viral neutralisation test

F

Although the neutralisation tests most commonly performed employ titration of serum mixed with a constant dosage of virus, the reciprocal technique (virus titrated-serum constant) is sometimes preferred. Either method is compatible with the micro system.

If desired, a two-dimensional assay (chessboard titration) can be performed easily on a single plate by making dilutions of virus or serum in either of these substances. This may be useful in determining the optimal combining proportion of antisera and antigen.

Also, a virus yield-reduction technique (Fig. 3.12), which determines the amount of residual virus remaining after neutralisation, can be easily accomplished in microplates. In this method, virus and antisera are mixed and

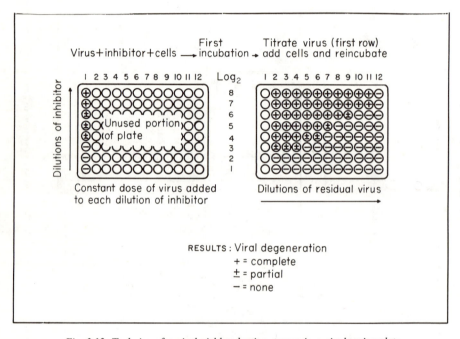

Fig. 3.12. Technique for viral yield reduction assays in a single microplate

incubated with cells in one portion of a plate. After a suitable incubation period, these are scored for the degree of cytopathic effect (unneutralised virus). Diluent is added to the unused portion of the plate and transfer of material is made from the previously infected cultures. Dilutions are made by loop, and fresh cells are added to these wells. The entire plate is reincubated and the amount of viable virus remaining after the neutralisation test is recorded. Such a procedure is more sensitive and quantitative than the ordinary neutralisation test and may be advantageous in interferon (IF) assays where low levels of active material may be present.

The isolation in microplates of viruses from clinical specimens has been accomplished with results agreeing favourably with tube isolations (unpublished observations). The toxicity that accompanies such specimens presents a definite problem, but can be overcome in most instances by the choice of collecting media, addition of a second aliquot of cells 24 hours after inoculation, or by subculturing.

The procedure for isolation is as follows: to a sterile plate with 'U' or flat bottomed wells is added 1 drop of diluent, 2 drops of specimen and 1 drop of cells to each well. Eight wells per specimen are equivalent to the inocula usually used in duplicate tissue culture tubes. Wells with cell controls are seeded in the rows between specimens enabling a plate to accommodate 6 specimens and 6 cell control rows. Plates are covered with sterile plastic covers (Linbro) and incubated in a well-humidified incubator under CO_2 tension (1–2%) to maintain a desirable pH. If in 24 hours a microscopic examination reveals cell toxicity, an additional drop of cells is added. These will overgrow the original cells clumped by the toxic substances. A passage may be indicated if toxicity persists. Plates are observed and read for CPE for a period of 6–8 days without a change of media. Subsequent passages are made by disrupting the cell sheet in all the wells with the tip of a dropper and drawing up the contents of all wells. Two drops are added to 1 drop of diluent in a new plate. Cells are added and incubation carried out as before. Remaining passage material may be frozen for future use. Specimens showing 4+ CPE are typed in microplates by adding 1 drop isolate, 1 drop typing sera and preincubation of the mixture for at least 1 hour at 35°C. Following this, cells are added and the plates are reincubated for some predetermined period (2–7 days).

Extreme care must be maintained in handling plates. If contents of wells are allowed to spatter on the cover due to rough handling, cross-contamination can result. Obviously, the careless handling of specimens containing unknown agents under such open conditions may be hazardous to the operator.

Thus far, adeno, polio, herpes, ECHO and influenza viruses have been isolated by this method using primary, semi-continuous strains (diploid lung), or continuous cell lines.

Requirement of special cultivation conditions such as the rolling of inoculated tubes for the isolation of rhinoviruses, would render this method impractical.

Besides the obvious advantages of this method of isolation, CPE often develops earlier and identification of viruses can often be accomplished in 4–6 days.

B. EVALUATION OF TESTS

Most conventional viral tests in tissue culture tubes employ either cytopathology, inhibition of cellular metabolism, erythrocyte agglutination or specific stains to indicate the propagation of virus or its inhibition. Assay endpoints are evaluated by either the micro or macroscopic appearance of the cells or by changes in appropriate indicators. These same techniques and methods of evaluation can be employed in the microplate system.

1. CYTOPATHOLOGY

Viral-induced degeneration in unstained viable cells can be viewed with the inverted microscope. Under low-power magnification ($35\times$), all of the cell monolayer can be seen. This enables the viewer to rapidly make an integrated appraisal without having to judge several microscopic fields, as is done with tube-cell cultures. Furthermore, 'A–B' comparisons of control and test results can easily be done by sliding the plate to and fro on the microscope stage. Thus, the observer can immediately assess the relative amounts of cell degeneration between the two wells before his mental image fades. Thus, the evaluation of microplate cell cultures is less subjective and tiresome than by conventional methods. At higher magnifications the characteristic cell degeneration caused by different groups of viruses is easily distinguished. Cells exhibiting various types of cytopathology are shown in Fig. 3.13.

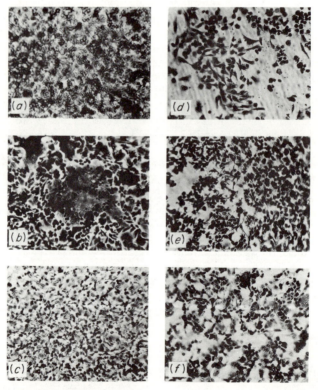

Fig. 3.13. Appearance of microplate cultures of normal and virus-infected HEp-2 cells stained with haematoxylin and eosin ($20\times$)
(a) *Uninfected cells*
(b) *Respiratory syncytial virus*
(c) *Poliovirus, type 3*
(d) *Herpesvirus*
(e) *Vaccinia virus*
(f) *Adenovirus, type 7*

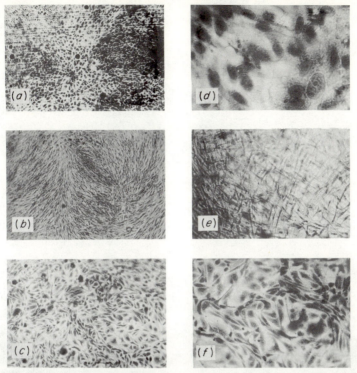

Fig. 3.14. Stained cell monolayers in microplates at various magnifications (all reduced by ½ on reproduction)
(a) Monkey kidney (primary)—May-Grünwald stain (35×)
(b) Human diploid lung—Wright stain (35×)
(c) Human embryonic kidney—May-Grünwald stain (100×).
(d) Monkey kidney—May-Grünwald stain (500×).
(e) Human diploid lung—Wright stain (100×).
(f) Human embryonic kidney—May-Grünwald stain (200×)
(From Sullivan and Rosenbaum[9]*, courtesy of the publisher of* Am. J. Epidem.)

If desirable, the cells can be fixed with a fixative that is not a solvent for the plastic and stained directly in the wells (Fig. 3.14). Thus, either microscopic cytological evaluation can be made or the entire residual cell sheet can be appraised macroscopically (Fig. 3.15).

2. METABOLIC-INHIBITION

The metabolic-inhibition test is perhaps the simplest way to evaluate test results[2-4] and is easily applied to the microtitre systems[2, 23]. An illustration of a metabolic-inhibition test for rhinoviruses is shown in Fig. 3.16. Lennette[4] has devised a specific media for such tests although the routine media described in

Neutralising antibody tests for rhinovirus type 2
in Giemsa stained HeLa cells

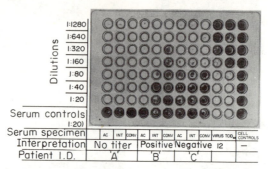

Serum specimen	AC	INT	CONV	AC	INT	CONV	AC	INT	CONV	VIRUS TCD	CELL CONTROLS
Interpretation	No titer			Positive			Negative			12	—
Patient I.D.	'A'			'B'			'C'				

Fig. 3.15. Macroscopic appearance of stained residual cell monolayers after a viral neutralisation assay ($\frac{1}{3}\times$ actual size). (From Sullivan and Rosenbaum[9], courtesy of the publisher of Am. J. Epidem.)

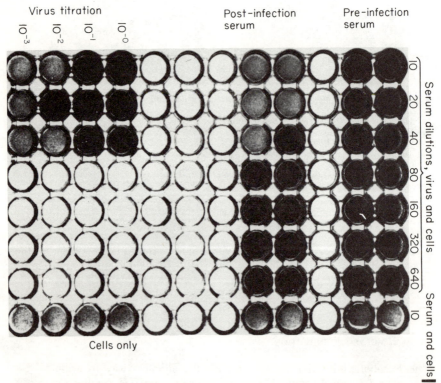

Fig. 3.16. Appearance of a metabolic inhibition test for rhinovirus infectivity and neutralisation serology (dark cups indicate metabolic inhibition). (From Stott and Tyrrell[23], courtesy of the authors and the publishers of Arch. Virusforsch.)

the procedure section can also be used. The disadvantage of the metabolic-inhibition method of assay is that tests must be incubated for a longer period to enable full development of colour to sharpen end-points, and also, many virus systems do not develop colour changes which can be consistently recognised. In contrast, direct microscopic viewing of microplate cells is very rapid and can indicate subtle changes in cellular morphology not detectable by gross-colorimetric examination.

C. INTERFERON ASSAYS

Interferon determinations in microplate have been described[16]. These can be carried out in a manner similar to that of tube cultures. Briefly, established monolayers are overlayed with fluids to be assayed. These cultures are incubated for 24 hours, fluids removed, and cells challenged with an appropriate dose of interferon-sensitive virus (Sindbis, ECHO-11). Significant inhibition of viral infection is indicative of interferon or interferon-like substances.

D. HAEMADSORPTION-INHIBITION

Certain viruses such as the myxoviruses may produce little or no cell damage in their replication. Vogel and Shelekov[32] have devised the procedure of applying erythrocyte suspensions to virus-infected cells and then examining the cultures microscopically for haemadsorption. Several investigators[24, 33] have shown this technique to be applicable for the titration of influenza and the parainfluenza virus group in the micro system. Wulff et al.[25] have reported a macroscopic technique for evaluating such tests. We have also used the macroscopic method with myxoviruses and feel that results are comparable to the more difficult microscopic procedure. The appearance of either the micro or macro-viral haemadsorption is shown in Fig. 3.17 and 3.18, respectively.

The procedure for myxovirus or other haemadsorption virus 'neutralisation' test is as follows. After preincubation of the virus-serum mixture in microplate wells, seed cells of secondary monkey kidney are added (see Procedures). The test is then incubated at 36°C for 3–5 days. Supernatant fluids are removed by decanting and 1 drop (0·025 ml) of 0·2% guinea pig (or chick) erythrocytes is added to each well. The plates are incubated at 4°C for 20 minutes.

The plates are then returned to room temperature, placed in a nearly vertical position and evaluated when control cultures (without virus) show 'streaking' of red cells (usually 10–30 minutes). Those cultures with mostly unadsorbed red cells will 'run' forming streaks of cells, indicating viral inhibition. In cultures where viruses have not been neutralised, the erythrocytes will be adsorbed to the cell monolayer (Fig. 3.18). Although a microscopic evaluation can be made and may be a more sensitive method to detect virus, it is more difficult to evaluate and reproduce.

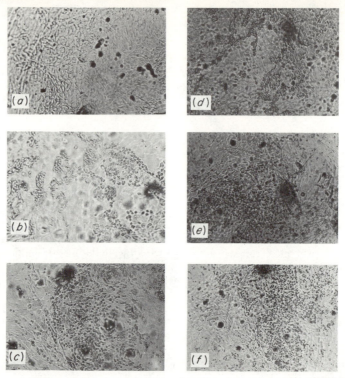

Fig. 3.17. Microscopic appearance of various types of viral haemadsorption in monkey kidney cell monolayers in microplates. (From Sullivan and Rosenbaum[9], courtesy of the publishers of Am. J. Epidem.)
(a) Uninfected control cells (100×)
(b) Vaccinia virus with chicken erythrocytes (100×)
(c) Parainfluenza, type 1 with guinea pig erythrocytes (100×)
(d) Vaccinia virus with chicken erythrocytes (50×)
(e) Influenza virus (A_2/Japan/1957) with guinea pig erythrocytes (50×)
(f) Parainfluenza, type 3 with guinea pig erythrocytes (50×)
(All reduced by $\frac{1}{2}$ on reproduction)

E. SPECIAL CYTOLOGICAL TECHNIQUES

Almost any type of conventional staining procedure for cells may be used for microplate monolayers provided the fixative is not a solvent for the plastic (acetone). Alcohol, formalin, osmium tetroxide, as well as multistep preparations, such as Carnoy and Bouin's fixatives, have been used with excellent results. Stains, such as methylene blue, methyl violet, haematoxylin and eosin, Giemsa, Feulgen, Wright and May-Grünwald reagents have all produced satisfactory cell preparations (Fig. 3.14).

All fixation, staining and washing procedures can be carried out in the plates, thus eliminating additional receptacles and minimising the wastage of reagents. Since cells are usually dehydrated during the staining procedures, a drop of oil

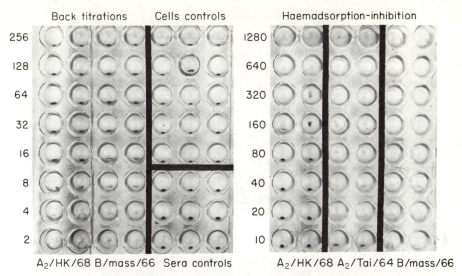

Back titrations   Cells controls   Haemadsorption-inhibition

256   1280
128   640
64   320
32   160
16   80
8   40
4   20
2   10

A₂/HK/68 B/mass/66 Sera controls   A₂/HK/68 A₂/Tai/64 B/mass/66

Fig. 3.18. Macroscopic appearance of influenza viruses (A₂/HK/68 and B/Mass/66), haemad-sorption titration (left plate) and typing (A₂/HK/68) with hyperimmune chicken sera (right plate). Cups with erythrocyte 'buttons' indicate lack of haemadsorption-inhibition by type-specific immune sera (right plate)

will return them to their normal shape. The stained monolayers can be examined with conventional microscopes by inverting the plate so the bottom surface is upward. Oil can be applied directly to the outer surface for high-magnification work.

Intracellular structural detail can be distinguished with the aid of either high power (dry) or oil immersion objectives. For such high resolution microscopy, the flat-bottom plates (see Equipment) are recommended since their planar characteristics minimise peripheral distortions.

D. FLUORESCENT ANTIBODY

The micro technique is useful for fluorescent antibody work (Fig. 3.19), as well as with non-conjugated fluorescent stains, such as acridine orange or coriphosphine-O.

When dark field condensers are employed for this work, it is necessary to place an oil column between the condenser and the material being viewed. Often the intact plate cannot be used. This problem can be remedied by cutting out the plastic (soft type) bottom of the well (containing the stained cells) with a cork borer with a diameter slightly smaller than the well. Gently preheating the utensil aids in achieving a smooth disc-like cut. The disc can be handled as a coverslip and mounted on a glass slide. High resolution fluorescent microscopy can be achieved in plates less than 1 mm thick.

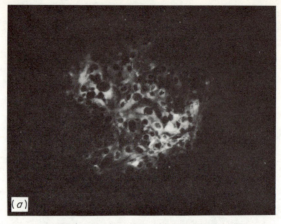

Uninfected monkey kidney cells

A₂ intluenza infected monkey kidney cells

Fig. 3.19. Normal and infected monkey kidney cell cultures stained with fluorescein-conjugated antibody to A_2/Hong Kong/1968 influenza virus ($\times$ c. 266).
(a) Uninfected monkey kidney cells
(b) A_2 influenza infected monkey kidney cells

ELECTRON MICROSCOPY

Rosenbaum *et al.*[34] have described an application of microplate cell cultures for electron microscopy. We will not attempt to explain all the details, such as choice of fixative, preparation of embedding media, and their degrees of hardness since these are a matter of personal preference. We shall merely point out how the use of microplates can facilitate such procedures (Fig. 3.20).

Ordinarily, preparation of cell monolayers for ultra-thin sectioning for electron microscopy can be an exasperating process. The cells usually have to be grown on a coverslip in a culture tube, removed from the coverslip surface and transferred to a plastic capsule before the fixation, dehydration, and embedding procedures can begin. The entire process, including preparation for sectioning, can be carried out in the microplate well. This eliminates most of the preliminary steps and provides many replicate specimens. Briefly, the procedures are as follows: after cells are established, the media is removed from the micro cultures and cells are washed with several rinses of 0·01 M phosphate-buffered saline (pH 7·0). Fixation is carried out in a well-ventilated hood and

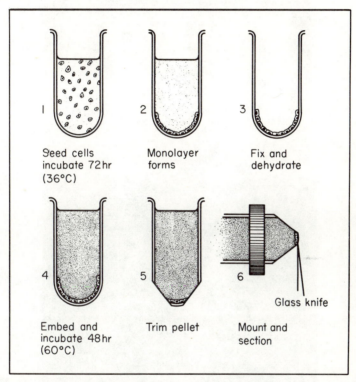

1 Seed cells
   incubate 72hr
   (36°C)

2 Monolayer
   forms

3 Fix and
   dehydrate

4 Embed and
   incubate 48hr
   (60°C)

5 Trim pellet

6 Mount and
   section

   Glass knife

Fig. 3.20. Preparation of microplate cell monolayers for embedding and ultra microtomy. (From Rosenbaum et al.[34], courtesy of the publisher of Lab. Pract.)

accomplished by placing the inverted plate in a cover lid containing filter paper saturated with 2% osmium-tetroxide. The culture plate is left in these vapours for 15 minutes. Then the fixed cells are rinsed with water and dehydrated with successive washes of increasing concentrations of ethyl alcohol (50, 70, 90 and 100%; 2 minutes each). The plates are air dried and the desired embedding agent (Epon, Araldite) is poured into the well and the infiltration and hardening period is carried out at 60°C for 48 hours. The hardened pellets can be cut out from the plates with the plastic jackets intact, or they can be removed by slitting and peeling (Fig. 3.21). These pellets are of a size convenient for several types

Fig. 3.21. Appearance of resin-filled microplate and solidified pellets for ultra microtomy. (From Rosenbaum et al.[34], *courtesy of the publisher of* Lab. Pract.)

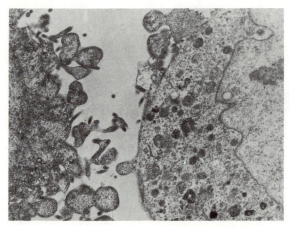

Fig. 3.22. Electron micrograph of an ultra thin section of HEp-2 cells cultured, fixed and embedded in microplates (× c. 9350). (From Rosenbaum et al.[34], *courtesy of the publisher of* Lab. Pract.)

of specimen holders and can be shaped and sectioned in the usual manner. Since the cell specimens usually are only one monolayer thick and are oriented in one plane, care must be taken to properly align the specimen surface with the knife edge to avoid making too many unnecessary cuts. Sections of reasonably good quality and detail can be obtained in the hands of experienced operators (Fig 3.22).

H. PHARMACOLOGICAL USES OF MICROPLATE SYSTEM

Many non-viable substances produce a lethal or toxic effect on cell cultures. These include biological products such as toxins, as well as a variety of drugs. Tissue cultures have been used to assay toxins from diphtheria[35] or gram-negative organisms (Sullivan, unpublished observations). Also, toxicity of vascular fluids, such as serum and plasma[36, 37], have been studied by tissue culture methods. Since many antitoxins can inhibit this effect, their potency can also be ascertained in tissue cultures[38].

Moreover, it may be important to screen antimicrobial drugs for toxicity, as well as specific effect, before elaborate animal programmes are initiated. The microplate cell system can be extremely valuable for such purposes, and mass screening of many compounds against a variety of cells can be accomplished without an extensive investment. Environmental conditions, such as temperature, atmospheric gas content, pressure or cell culture media, can be easily varied.

The microplate technique is very suitable for antiviral assays, whereby several compounds can be tested against a number of viruses simultaneously.

The procedure for the above tests is quite similar to those described for viral or neutralisation assays.

POTENTIAL USES

Although none of the following microplate techniques have as yet been actually performed, they appear to have excellent prospects for being useful. In most instances, they only represent modifications of methods which employ tubes or larger receptacles.

A. IMMUNOLOGICAL

The establishment of leucocyte cultures in microplates has been reported[39]. Such cultures could be useful for studies on lymphoblast formation, genetic comparisons, interferon production, *in vitro* histocompatibility tests or for the determination of potency of antilymphocyte serum. Many other immunological techniques could also be employed by similar procedures. Tests for blood-group compatibility, leucoagglutinins and leucocytotoxins could easily be adapted to the microplate. By setting up a chessboard pattern, mass histocompatibility tests could be rapidly performed. The determination of the cytolytic activity of non-viral antigen/antibody complexes and the role which complement may play in such activity, are other examples of how microplate cultures may be utilised.

B. CELL CLONING

Cell cloning may be easily accomplished by the combination of microplates and loop diluters. Seed cells can be added to the plate at a given concentration.

By successive transfers of such seed suspensions, a dilution containing only a few or one cell can be rapidly achieved with many similar replicate dilutions. The presence of a single cell in a well can be confirmed by microscopic examination and the plate incubated. Thus, the progeny of this cell can be transferred to a larger vessel to provide a cloned cell population.

C. AUTOMATION

Perhaps the greatest potential use of the microtitre system is in its adaptability for automation. A schematic conception of such a process is shown in Fig 3.23. It was previously mentioned that commercial automatic diluters are available.

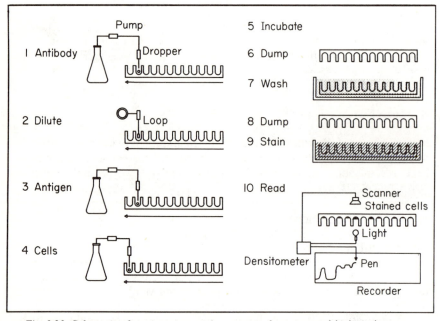

Fig. 3.23. Schematic of a micro tissue culture system for automated biological assays

This device replaces the operator in performing mechanical functions of addition of reagents and diluting. By integrating this system with another for staining residual cells (Fig. 3.15), and measuring these with a densitometer, a graphic record of the assay could be produced.

It is recognised that such procedures, at present, may have only limited application, but they are entirely feasible without a great deal of engineering obstacles and may prove useful in the future.

TECHNICAL PROBLEMS

A. NON-SPECIFIC TOXICITY

The greatest problem encountered with the micro-tissue culture system is non-specific cytotoxicity. Although this may be due to a variety of causes, the most common are either inherent plate properties or unknown factors in serum.

The exact cause of plate toxicity is not known, but is believed to be due to the moulding process in which the die is covered with an agent to facilitate release of the plastic forms. This problem can usually be corrected by alcohol washes. Occasionally, the toxicity of some batches of plates cannot be remedied by this procedure and must be discarded. This is probably due to the composition of the plastic itself. Consultation with the plate manufacturer has helped to identify the source of such infrequent problems.

Serum cytotoxicity is a problem which needs further investigation. Such effect is also observed in conventional tube cultures, but because of the greater dilution factor, it has not been as bothersome. Failure to separate serum from the clot soon after coagulation has been completed tends to increase the severity of the toxicity (Sullivan and Rosenbaum, unpublished observation). Refrigeration of clotted blood soon after collection, however, retards toxicity. Also, repeated thawing and freezing of serum tends to increase toxicity. Techniques used to reduce serum toxicity include heat-inactivation (56°C for 30 minutes), or absorption with tissue or organ powders.

B. PLATE IRREGULARITIES

At various times, plates may be defective due to score marks which appear on the well bottom. These marks interfere with the establishment of a confluent monolayer and may make microscopic observation difficult. The marks are usually due to mould irregularities, and the manufacturer should be advised to replace them.

C. HAZARDS

Perhaps the most serious problem which may be unrecognised is that a potential insidious hazard exists if plates with infected tissue cultures are handled too casually. This is easily done because of the routine nature which develops from the simplicity of the system itself. All precautions should be taken to impress even experienced operators of this danger, and all plate tissue cultures should be handled carefully and disposed of by soaking in 0·5% HCl or autoclaved.

CONCLUSION

Despite these problems, the microplate system can be useful for a variety of procedures which involve tissue cultures or agents which propagate in them or produce certain effects. There are probably many areas of application which have not been mentioned in the preceding discussion. Such uses are only limited by the ingenuity of the investigator. The procedures suggested herein are only guides and may be modified to suit individual needs. perhaps the greatest advantage of the microplate system is that it liberates the investigator from the tediousness, expense and routine of the tissue culture tube and permits him more time for creative research.

ACKNOWLEDGEMENTS

The authors wish to thank Robert O. Peckinpaugh, CAPT, MC, USN, Commanding Officer, NAMRU-4, for the support given to them over the past years for the development of the micro method. We are also indebted to Mr. C. E. Knight, Chief of Publications Division, NAMRU-4, for the excellent technical assistance rendered in the preparation of this article.

From Research Project MF 022.03.07-4003. The opinions and assertions expressed herein are those of the authors and cannot be construed as reflecting the views of the U.S. Navy Department, or of the U.S. Naval Service at large. Usage of a commercially available product in connection with this study cannot be construed as an endorsement of such product. The experiments reported herein were conducted according to the principles enunciated in 'Guide for Laboratory Facilities and Care' prepared by the Committee on the Guide for Laboratory Animal Resources, National Academy of Sciences—National Research Council.

REFERENCES

1. BARSKI, G. and LEPINE, P., *Annls Inst. Pasteur, Paris*, **86**, 693 (1954)
2. MELNICK, J. L. and OPTON, E. M., *Bull. Wld Hlth. Org.*, **14**, 129 (1956)
3. JOHNSTON, P. B., GRAYSTON, J. T. and LOOSLI, C. G., *Proc. Soc. exp. Biol. Med.*, **94**, 338 (1957)
4. LENNETTE, E. H., NEFF, B. J. and FOX, V. L., *Am. J. Hyg.*, **65**, 94 (1957)
5. RIGHTSEL, W. A., SCHULTZ, P., MUETHING, D. and MCLEAN, I. W., *J. Immun.*, **76**, 464 (1956)
6. TAKATSY, G., *Acta microbiol. hung.*, **3**, 191 (1955)
7. SEVER, J. L., *J. Immun.*, **88**, 320 (1962)
8. ROSENBAUM, M. J., PHILLIPS, I. A., SULLIVAN, E. J., EDWARDS, E. A. and MILLER, L. F., *Proc. Soc. exp. Biol. Med.*, **113**, 224 (1963)
9. SULLIVAN, E. J. and ROSENBAUM, M. J., *Am. J. Epidem.*, **85**, 424 (1967)
10. EDWARDS, E. A. and PECK, R., *Rep. Bur. Med. Surg.*, U.S. Navy Dept., Washington, D.C. 1964, No. MR 005.09-1300.1
11. EAGLE, H., *Science, N.Y.*, **122**, 501 (1955)
12. EARLE, W. R., *J. natn Cancer Inst.*, **4**, 165 (1943)
13. MORGAN, J. F., MORTON, H. J. and PARKER, R. C., *Proc. Soc. exp. Biol. Med.*, **73**, 1 (1950)
14. LEIBOVITZ, A., *Am. J. Hyg.*, **78**, 173 (1963)

15. SCHMIDT, N. J., LENNETTE, E. H. and HANAHOE, M. F., *Proc. Soc. exp. Biol. Med.*, **121**, 1268 (1966)
16. TILLES, J. G. and FINLAND, M., *Appl. Microbiol.*, **16**, 1706 (1968)
17. YOUNGNER, J. S., *Proc. Soc. exp. Biol. Med.*, **85**, 202 (1954)
18. HAYFLICK, L. and MOOREHEAD, P. S., *Expl Cell Res.*, **25**, 585 (1961)
19. LAMB, G. A. K., GLEZEN, W. P. and CHIN, T. D. Y., *Publ. Hlth. Rep., Wash.*, **80**, 463 (1965)
20. KENDE, M. and ROBBINS, M. L., *Appl. Microbiol.*, **13**, 1026 (1965)
21. SCHMIDT, N. J., LENNETTE, E. H. and DENNIS, J., *J. Immun.*, **100**, 99 (1968)
22. GWALTNEY, J. M., *Proc. Soc. exp. Biol. Med.*, **122**, 1137 (1966)
23. STOTT, E. J. and TYRRELL, D. A. J., *Arch. Virusforsch.*, **23**, 236 (1968)
24. SCHMIDT, N. J., LENNETTE, E. H. and HANAHOE, M. F., *Proc. Soc. exp. Biol. Med.*, **122**, 1062 (1966)
25. WULFF, H., SOEKEN, J., POLAND, J. D. and CHIN, T. D. Y., *Proc. Soc. exp. Biol. Med.*, **125**, 1045 (1967)
26. HARRIS, D. J., WULFF, H., RAY, C. G., POLAND, J. D., CHIN, T. D. Y. and WENNER H. A *Am. J. Epidem.*, **87**, 419 (1968)
27. MOREAU, P. and FURESZ, J., *Can. J. Microbiol.*, **13**, 313 (1967)
28. KRIEL, R. L., WULFF, H. and CHIN, T. D. Y., *Proc. Soc. expl. Biol. Med.*, **130**, 107 (1969)
29. ROSENBAUM, M. J. Unpublished observations
30. SCHMIDT, N. J., LENNETTE, E. H. and KING, C. J., *J. Immun.*, **97**, 64 (1966)
31. ROSENBAUM, M. J., DE BERRY, P., SULLIVAN, E. J., EDWARDS, E. A., PIERCE, W. E., MULDOON, R. L., JACKSON, G. G. and PECKINPAUGH, R. O., *Am. J. Epidem.*, **88**, 45 (1968)
32. VOGEL, J. and SHELEKOV, A., *Science, N.Y.*, **126**, 358 (1957)
33. SMITH, C. B., CANCHOLA, J. and CHANOCK, R. M., *Proc. Soc. exp. Biol. Med.*, **124**, 4 (1967)
34. ROSENBAUM, M. J., EARLE, A. M., CHAPMAN, A. L. and SULLIVAN, E. J., *Lab. Pract.*, **17**, 713 (1968)
35. LENNOX, E. S. and KAPLAN, A. S., *Proc. Soc. exp. Biol. Med.*, **95**, 700 (1957)
36. ROSENBAUM, M. J., MILLER, L. F., SULLIVAN, B. and ROSENTHAL, S. R., *Fedn Proc. Fedn Am. Socs exp. Biol.*, **19**, 357 (1960)
37. BENO, D. W., LYTLE, R. I. and EDWARDS, E. A., *Bact. Proc. (Am. Soc. Microbiol.)* p. 120 (1965)
38. SOUSA, C. P. and EVANS, D. G., *Br. J. exp. Path.*, **38**, 644 (1957)
39. BRODY, J. A. and HUNTLEY, R., *Nature, Lond.*, **208**, 1232 (1965)

FOUR

Tissue culture in diagnostic virology

JOHN BERTRAND

Department of Clinical Virology, St Thomas's Hospital, London

Tissue culture monolayers have been in use for various purposes in the laboratory for some years. In 1949 the first report of the propagation of a virus in this type of culture was made by Enders, Weller and Robbins[1]. A method of isolating and propagating poliovirus was described, and this became a model for further work in attempting to isolate viruses from various sources. Mono-layer tissue cultures have since become a corner stone in the structure of modern systematic virology.

This chapter will give a brief history of tissue cultures in virology, followed by an account of the uses and abuses of the cultures in the diagnosis of viral infections in man. Reference will be made in each section to observations in tissue culture applied to virology that have been reported in recent years, together with some comments based on the author's experience in isolating viruses from patients.

HISTORY

Enders' work on poliovirus was followed by the isolation of many hitherto unknown viruses, among them the coxsackie viruses[2, 3], the ECHO viruses[4], the adenoviruses[5, 6] and, more recently, respiratory syncytial virus[7] and cytomegalovirus[8]. In addition many known viruses were found to multiply in monolayer cultures[9-11], examples including herpes simplex and influenza viruses.

Improvements in techniques that went beyond the observation of visible changes in the tissue culture led to the discovery of the haemadsorbing viruses of the myxovirus group[12]. Observations of the phenomemon of interference resulted in one of the first isolations of rubella virus[13].

At the time of writing the number of new viruses isolated in monolayer cultures has been drastically reduced compared to the period immediately after the isolation of poliovirus. Some viruses have proved extremely difficult to isolate in monolayer cultures and more sophisticated methods such as the use of organ cultures is of rapidly increasing importance in this field[14]. The importance of the monolayer culture is not, however, diminishing. Had their use never been described a large number of severe diseases of man and animals would remain a mystery, and protection of susceptible human or animal patients would often have been impossible.

THE DIAGNOSTIC VIRUS LABORATORY

This is a laboratory organised for the purpose of demonstrating viral infections in patients by using isolation or serological techniques. A diagnostic virus laboratory should be able to demonstrate infection by as many of the common viruses as possible. with methods involving the use of fertile eggs. laboratory animals, serology and tissue cultures. It is with the tissue cultures and the techniques associated with them that this chapter is concerned.

Tissue cultures have three uses in diagnostic virology:

1. The isolation and identification of viruses.
2. The demonstration of viral infection by showing significant rises in antibody levels in paired sera.
3. The preparation of antigens and antisera for use in serologic tests.

The three main sources of tissue for monolayer cultures are:

1. Animal tissue, e.g. monkey kidneys.
2. Human malignant tissue from which the HeLa cell and similar cell types were derived.
3. Human foetal tissue.

EQUIPMENT AND MATERIALS

The basic requirements of a department that uses tissue culture methods are similar to the needs of many other biological laboratories and may be obtained from similar sources. Equipment such as incubators, water baths and small centrifuges are required. The author's department possesses a walk-in hot room. This room is fitted with solid metal shelves, and a temperature gradient has been established in it by means of carefully sited fans. This arrangement is useful as some viruses multiply optimally at 33–35°C, whereas others multiply optimally at 37°C. A roller machine is essential for the isolation of some respiratory viruses. Rolling of tissue cultures at a speed of 8–10 revolutions per hour, whether or not they are inoculated, is frequently beneficial for the cultures.

Space should be set aside in a clean, well ventilated room for carrying out all procedures associated with the tissue cultures. Viruses should never be taken into this room. The room should be fitted with ultra-violet striplights that can be operated to within six inches of the bench surface. These lights should be switched on at the end of each period of work to ensure that the working surface remains free of organisms. Lights fitted high above the bench are not sufficient, as most of the ultra-violet light is absorbed before it reaches the bench. It should be remembered that ultra-violet light can cause burns to the skin and the cornea.

Glassware for tissue culture is very simple, and consists of bottles and tubes of suitable size. All glassware must be clean and sterile. A suitable tube size is 100 mm × 10 mm. These tubes are received in a clean enough state to be sterilised without washing. The author finds it more convenient to discard these tubes after one use rather than wash and re-use them. The best type of seal for these tubes is the silicone rubber bung, which is almost indestructible. These represent a high initial cost and an excellent substitute is the white rubber bung that has a life of about six months.

The range of medical flat bottles is quite suitable for keeping stock cultures. Roux bottles may be used when optically clear glass is required. These bottles, and other glassware, are too expensive to discard after one use and a washing-up method must be devised. There are many methods available, some of which are quite complicated. The one described is that currently used by the author's department.

Glassware is sterilised by autoclaving before the contents of the bottles are emptied. Bottles are then filled with a dilute solution of the detergent Pyroneg. After overnight soaking the inner surface is vigorously brushed to remove any cells that may be adhering. The bottles are then rinsed at least six times in tap water. They are then finally rinsed in distilled water. The bottles are dried by gentle heat, capped, and sterilised by autoclaving. Bottle caps and rubber liners are boiled in the detergent together with the rubber bungs. These are then rinsed in the same way as the bottles. There is a constant control on all the glassware, the tissue cultures themselves. Should the cultures fail to thrive, one should ensure that the washing-up process is being used properly.

Plastic ware for tissue culture is readily available, though it tends to be expensive: Petri dishes in clear plastic have been available for some years, but it is only recently that they have been produced specifically for tissue culture. The bottle and tubes now produced are of excellent optical quality. Baby's feeding bottles used for growing cells for viral antigens and for growing cells in bulk are useful, although the plastic is of poor optical quality.

CULTURE MEDIUM

A number of commercial firms produce tissue culture media of good quality. It is usually more convenient, and definitely cheaper, to purchase these media in concentrated form. T.C. medium 199[15] and Eagle's minimum essential medium[16] are probably the most useful fluids to have in the laboratory, supplied at ten times the concentration for use. Most cell lines will grow on either of these media supplemented with 5–10% serum and buffered to pH 6·8–7·2.

4·4% sodium bicarbonate solution adjusted to pH 7·0 with carbon dioxide is used for buffering media to the required pH.

Serum used to supplement media is usually calf serum. Foetal bovine serum is often used, particularly to maintain cultures, and a small amount of rabbit serum may be required. There are a number of commercial sources of these sera, some being more reliable than others. Most sources supply material that has been fully tested for sterility and for its ability to support the growth of the common cell lines. Many suppliers of serum will send samples of large batches to the laboratory for testing under the condition in which they will be used. Suitable batches will be reserved for dispatch to the user as required.

A number of other media are on the market in addition to those mentioned. It is not usually necessary for diagnostic laboratories to hold stocks of these.

Antibiotics are readily available from the hospital dispensary or from a commercial source. Penicillin and streptomycin are most commonly used, together with an antifungal agent; gentamycin at 50 units/ml is now being used by some laboratories instead of streptomycin as it is more stable and can be sterilised by autoclaving at 121°C for 15 minutes. It is convenient to keep the antibiotics in aliquots. One aliquot added to a batch of medium should give a final concentration of 100–200 units of penicillin and 100–200 µg of streptomycin per ml. Antifungal agents are added to give a final concentration of up to 25 µg per ml. Care should be taken not to add too much of any antibiotic, as the tissue cultures may be sensitive to them.

SELECTION OF CELL LINES

A recent development in diagnostic virology has been the appearance of small low-budget departments in both teaching hospitals and large general hospitals with a small laboratory staff. As resources and funds are often limited, these laboratories take care to select the best cell lines available for the current work. The systems recommended in the following pages are probably the most that a small department could administer during a normal week. In my own experience it is unnecessary for a small diagnostic laboratory to have more than three types of culture in regular use. The types recommended are:

1 Primary or secondary monkey kidney.
2. A human embryonic fibroblast line.
3. A continuous cell line of the HeLa type.

The sensitivity of these cell types is shown in *Table 4.1*.

The ideal cell system has yet to be developed. The cell systems chosen should, however, have between them the following list of attributes:

1. They should be sensitive to as wide a range of viruses as possible, optimally to all the known human types.
2. The cells must maintain their normal morphology for a minimum culture period of 14 days, unless they become infected with a virus. Should such an infection occur, the cells must show a clearly defined effect in a reasonable time.
3. The cells should be maintained with simple media supplemented with the available sera.
4. Cells must be amenable to storage in liquid nitrogen.
5. Cells must be free from adventitious agents such as moulds and viruses.

Table 4.1 RELATIVE SENSITIVITY OF THE COMMON TISSUE CULTURES

Virus	PMK	HeLa	Human embryo lung	Human amnion	Other
Poliovirus	+++	+++	+++	+++	–
ECHO	+++	+	+++	++	–
Coxsackie A	(+)	–	–	–	Suckling mice +++
Coxsackie B	+++	+	+++	++	Suckling mice +++
Rhinovirus M strain	+++	–	+++	–	–
Rhinovirus H strain	–	–	+++*	–	Organ cultures ++
Influenza group	+++	+	+	+	Fertile eggs +++
Parainfluenza group	+++	+	+	++	Fertile eggs +
Mumps	+++	+	+	++	Fertile eggs ++
Measles	+++	+	+	++	–
RSV	++	+++	–	–	–
Adenovirus	++	+++	+	++	–
Herpes simplex	–	++	+++	+++	Fertile eggs ++
Herpes zoster	–	–	+++	+++	–
Cytomegalovirus	–	–	+++	–	–
Vaccinia	+++	+++	+++	+++	Fertile eggs +++
Variola	+	+	+	+	Fertile eggs +++
Rubella	++	–	–	–	RK13 +++
Rabies	–	–	–	–	Suckling mice +++
LCM	–	–	–	–	Suckling mice +++
Arboviruses	++	–	–	–	Suckling mice +++

+++	Most useful system.	ECHO: Enteric cytopathic human orphan.
++	Best alternative to above.	RSV: Respiratory syncytial virus.
+	Variable isolation rate.	LCM: Lymphocyticchoriomeningitisvirus.
–	No growth.	PMK: Primary monkey kidney cells.
*	Only about 30% of H strain rhinoviruses will grow in this system.	

PRIMARY AND SECONDARY MONKEY KIDNEY

These cells are obtained from the freshly excised kidneys of various species of monkeys. Rhesus, patas, cynomolgus and vervet monkeys provide the common sources of kidneys.

The kidney is removed from the monkey using aseptic techniques and the cortex is then removed and chopped into small pieces. The pieces are treated with trypsin so that the epithelial cells are disaggregated. The suspension of cells is counted in a haemocytometer and is then transferred into a suitable growth medium at a concentration of 300 000 cells per ml. The growth medium recommended is medium 199 with 5% calf serum buffered to pH 7·2.

The suspension is inoculated into tubes and bottles as required, 1 ml being inoculated into each tube and the appropriate amount in bottles. Monolayers become confluent in 2–4 days. Confluent monolayers are maintained on serum-free medium 199, provided that this is changed at least twice weekly. Cultures used to isolate members of the myxovirus group are washed free of serum before inoculation to ensure that no viral inhibitors are present.

Secondary cells are obtained from the bottles that are seeded at the same time as the tubes. The monolayers are removed from the glass by means of a chelating agent or an enzyme. Cells are counted and distributed into tubes at a concentration of 100–300 000 cells per ml.

Secondary rhesus monkey kidney cells, obtained from the Medical Research Council laboratories at Hampstead, are received in plastic baby's feeding bottles. The cells are removed from the containers by using the method described. There are three advantages gained by using secondary cells:

1. More cells are obtained from each kidney, and thus less monkeys are needed.
2. The cells can be seen to be growing properly and are not contaminated with simian viruses.
3. It is possible to keep reserve bottles in the incubator in case of accidents when processing the chosen cells.

Monkey kidney cell cultures in diagnostic virology are in some ways analagous to the blood agar plate in bacteriology. The range of viruses supported is very wide and is shown in *Table 4.1*. The cells grow quickly to form monolayers, and provided that the medium is changed regularly they will maintain for up to four weeks in good condition. Monkey kidney cells can be stored in liquid nitrogen and recovery rates are excellent.

Primary monkey kidney cells, and to a lesser extent secondary cells, harbour simian viruses which can interfere with isolation work. Among the commoner of these contaminants is simian virus 5, which is a member of the paramyxovirus group. Cells infected with this virus exhibit the phenomenon of haemadsorbtion. Presence of simian virus 5 in the cultures will interfere with attempts to isolate most of the members of the myxovirus group. Addition of an antiserum to this virus to the tissue culture fluids will help to eliminate this common problem. Other common contaminants inclue simian virus 40 vacuolating agent, which causes spontaneous degeneration and has been shown to be oncogenic, and the so-called foamy agent, causing the development of vast vacuolated syncytia (Fig. 4.1). The development of cytopathic effects produced by some of these viruses may be retarded by changing the medium daily. However, the most suitable remedy for this type of contaminant is to discard the cultures and await a batch of cells that are free from viruses.

There are some dangerous contaminants to be found in monkey cells, the most infamous being Herpes virus simiae (Virus B). This virus is the cause of mouth ulcers in monkeys. In man a severe and usually fatal encephalomyelitis is caused. A second cause of severe disease in man was isolated during 1968 in Germany[17]. A number of patients became ill with a hepatitis to which a number succumbed. The causative organism was traced back to the monkeys. This organism is now thought to be the prototype of a hitherto unknown group. It is very important that any person who injures himself on apparatus containing monkey cells, or is bitten, or scratched by a monkey should seek medical advice as soon as possible after the incident.

HeLa AND RELATED CELL TYPES

HeLa cells originated from cultures of a carcinoma of the cervix in 1952[18]. They

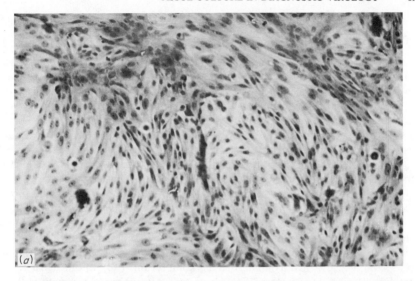

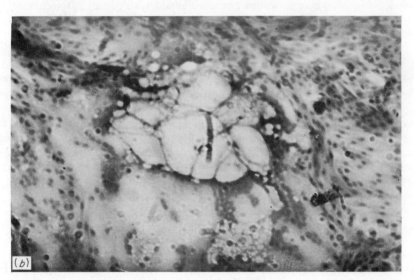

Fig. 4.1. (*a*) *Normal monkey kidney cells*
(*b*) *Vacuolating agent in monkey kidney cells*

are malignant cells with a variable karyotype and have a morphology similar to that of pavement epithelial cells. Related cell types found in the laboratory include HEp2[19] and KB[20]. All these cells are remarkably easy to maintain and use in the laboratory, they are sensitive to a wide range of viruses. The main use of these cells is to isolate and identify adenoviruses, Herpes simplex virus and respiratory syncytial virus.

HeLa cells can be propagated indefinitely and are kept in the laboratory as monolayers in bottles. The monolayers are disaggregated and redistributed twice weekly as follows. The cells are stripped from the glass using a chelating agent (Versene 0·02%) or an enzyme (0·25% trypsin). The cells are suspended in growth medium and counted. The suspension is then diluted in growth medium to give a concentration of 100 000 cells per ml. One medical flat bottle (100 cm³) containing a confluent monolayer will yield approximately 60 tubes at this seeding rate. Should cells be required quickly, no damage will result from seeding tubes at up to 500 000 cells per ml. Stock cultures are seeded at the same rate as the cultures in tubes.

A suitable growth medium for these cells is Eagle's minimum essential medium (M.E.M.) supplemented with 10% calf serum and buffered to pH 7·0. Confluent cultures may be maintained on M.E.M. with 2% calf serum or foetal calf serum, buffered to pH 7·4.

Cells that are required to support the growth of respiratory syncytial virus (RSV) need more careful treatment if they are to retain their sensitivity to this virus. The Bristol strain of HeLa cells is often used to isolate this virus. These cells should be grown and maintained on rabbit serum. The author's department employ HEp2 cells to isolate RSV. These cells will support the growth of RSV after being grown with serum and maintained with foetal calf serum. It is advisable to check the selected cell system for sensitivity to RSV at regular intervals.

HeLa cells can be stored in liquid nitrogen and recovery rates are excellent.

Most cell lines of the HeLa type suffer from chronic infection with Mycoplasma species. These organisms usually multiply slowly and do not interfere with virus growth. Occasionally, however, their rate of multiplication will increase, and the cells fail to thrive because of the excess acid produced by the contaminant. It may be possible to treat the cultures with antibiotics such as tetracycline. A more satisfactory remedy is to discard the affected cells and replace them with fresh cells from the storage system. Contamination of these cells with bacteria, moulds or viruses is rare provided the cells are handled with reasonable care.

FIBROBLAST CELLS

These cells are obtained from several sources, among them being chick embryos and the aborted human foetus. The cells are spindle shaped and grow in tight whorls said to resemble finger prints. The appearance of a normal monolayer is highly distinctive.

The tissue selected is often human embryonic lung (HEL). Other tissues have been successfully used, notably skin-muscle. The best known strain of fibroblasts is the Wistar Institute No. 38 (W.I. 38). These cells are sensitive to many viruses, including enteroviruses, all the Herpes group viruses, and to a large proportion of the rhinoviruses. Its main disadvantages are its poor growth on many types of Eagle's medium, and its poor growth on some batches of calf serum. It is more convenient to initiate one's own cell line from fresh lung.

Embryonic lungs should ideally come from 14–18 week embryos. In practice one accepts the tissue offered as, even under the new abortion laws, human embryos are not readily available. The lungs are freed from fibrous tissue and the remaining soft pale coloured tissue chopped into pieces about one cubic millimetre in size. The pieces are washed to free them from blood and are then suspended in a volume of 0·25% trypsin. This suspension is incubated at 37° C for 30 minutes after which the supernatent fluid is discarded. The remaining tissue is subjected to three further 30 minute treatments with trypsin, the supernatent fluid being collected after each treatment. The suspended cells are removed from the trypsin by centrifuging lightly. The button of cells is resuspended in approximately ×70 of their own volume of growth medium. The suspension is then placed in medical flat bottles to await the development of monolayers.

Typical monolayers may not be visible until 48–72 hours after seeding the bottles. The growth medium should be replaced with fresh medium every 24 hours for the first three days to remove any debris. Once typical confluent monolayers are obtained it should be possible to strip the cells from the glass with 0·25% trypsin and seed the cells into fresh containers in the same way as HeLa cells. Suitable growth and maintenance media for these cells are identical to those used to propagate and maintain HeLa cultures.

Once a new line is established it is necessary to test it with the viruses that it will be expected to support and under the conditions in which these viruses will be encountered. This will involve using clinical material in a cell line of known susceptibility in parallel with the new line. When the new line has been shown to isolate the same number of viruses at the same rate as the known cells without giving a higher proportion of false results, it may be used routinely.

Fibroblast lines are often referred to as semi-continuous, that is, they may be treated with an enzyme and redistributed a finite number of times. This number usually falls between 30 and 50 passages. In order to perpetuate the use of a cell line in the department, it is essential to store as much of the cells of low passage number as the system will allow. These cells are usually amenable to storage and recovery rates are good.

Monolayers are passed into fresh containers in the same way as HeLa cells. Cell suspensions are counted and seeded into tubes and bottles at 300 000 cells per ml. Confluent monolayers are obtained in three to four days. One medical flat bottle (100 cm³) will yield about 30 tubes at this seeding rate. Cultures that are carefully maintained by replacing the medium at least twice weekly can be kept in good condition for up to eight weeks.

The main use of these cells is to isolate and identify the viruses of the Herpes group, in particular the more fastidious members of the group such as Varicella-Zoster and cytomegalovirus. Human cytomegaloviruses cannot be isolated in any cells other than fibroblasts of human origin. This virus is rapidly increasing in importance in medicine due to its association with intrauterine infection, with the post-perfusion syndrome and with infectious mononucleosis-like illnesses, and for this reason fibroblasts are essential to the diagnostic laboratory. Fibroblasts are also excellent for the isolation of enteroviruses, particularly the ECHO group. Some lines will support the growth of H strain rhinoviruses.

Latent viruses may be found in fibroblast cultures. Workers should check that their lines are not carrying rubella virus. Infection of pregnant women with this virus during the first trimester of pregnancy results in about 30% of the foetus being infected in utero, the infection being accompanied by severe foetal damage. Rubella in the first trimester of pregnancy is taken as sufficient grounds for a therapeutic abortion. Tissues from these embryos may be found to be heavily infected with rubella virus.

HUMAN AMNION CULTURES

These cells are obtained from the amniotic membrane of placentae obtained at caesarian section. The amnion is removed from the placenta, and is washed free of blood. The clean tissue is cut into small pieces and treated with 0·25% trypsin. The cell suspension is washed free of trypsin, counted, and seeded into tubes at 500 000 cells per ml. Monolayers of epithelial cells develop in 4–7 days. These cells are excellent for the isolation of enteroviruses, particularly poliviruses, and for Herpes Varicella-Zoster virus. Their limitations are three-fold:

1. They are unpleasant to handle, being rather messy.
2. The placenta may arrive at an inconvenient moment.
3. They may fail to grow at all.

There is no virus that will grow in aminon cultures that will not grow as well in the tissues already discussed, which are altogether more convenient to administer.

MANAGEMENT OF CELL CULTURES

A diagnostic virus laboratory requires a constant supply of tissue cultures in good condition, and that are likely to last at least 14 days. Cultures must therefore be prepared as often as is conveniently possible, usually once a week. Rapidly growing lines such as HeLa cells should be redistributed twice weekly. Enough cultures in tubes should be produced to provide what, by experience, is found to be in excess of requirements for a normal week. Sufficient stock culture bottles should also be seeded to provide for the following week's cultures and emergency supplies.

It is convenient to put up one batch of monkey kidney and fibroblasts, and two batches of HEp2 each week. A large number of monkey kidney cells are inoculated for two reasons. Firstly, isolation of an enterovirus means the use of a large number of tubes for its identification by neutralisation. Secondly, should there be a flood of specimens it is inconvenient to remove the kidney from a monkey at short notice, and it will probably be impossible to obtain an immediate supply of secondary cells. To be forced to remove cells from the store for this reason is somewhat wasteful. It is therefore considered to be more economical to discard up to 100 unused tubes each week rather than to be short of cultures in an emergency. Shortage of cells may also necessitate freezing some specimens until cultures are available, a procedure that will inactivate a proportion of any viruses that may be present.

Care must be taken when handling the cell cultures to avoid the introduction of contaminants. If a special tissue culture room is not available it is better to deal with the cultures before any viruses are used. The bench should be clean and clear of any equipment that is not required during the tissue culture procedures. Tubes and bottles should be flamed on opening and before closing. Forceps should be used to pick bungs from their containers. These elementary precautions are often ignored in the presence of antibiotics and anti-fungal agents in the tissue culture fluids. Antibiotics are useful but not infallible, and it should be remembered that a tissue culture once contaminated is very difficult to recover.

LIQUID NITROGEN STORAGE SYSTEMS

Contamination of tissue cultures will occur even in the best organised departments. It is at times when the cell systems are failing for any reason that the liquid nitrogen store comes into its own. These stores are supplied by the Linde division of Union Carbide. The model shown in Fig. 4.2 accomodates a large number of ampoules which are immersed in the liquid nitrogen. Later models

Fig. 4.2. Liquid nitrogen container showing canister and canes in which ampoules are stored (Courtesy of Union Carbide)

enable the containers to be stored in the gas phase and not immersed in the liquid. This means that screw capped bottles such as the bijou bottle (7 cm³) can be used to store cells.

The method of storage is much the same for all the cell lines. Confluent young cultures of good morphological appearance are selected. Cells are stripped from the glass by the usual method, and are concentrated to at least 1 000 000 cells per ml preferably more. The storage medium consists of the normal growth medium for the cells with 10–15% of a preservative added. The preservative may be buffered glycerol or dimethyl sulphoxide. The cell suspension is placed in strong glass ampoules, taking care not to overfill them. The ampoules are carefully sealed. The contents are then frozen at a controlled rate of 1°C a minute from 4°C to below −50°C. Faster or slower freezing will cause the formation of ice crystals and result in much loss of viability in the culture. Apparatus is available with the storage flask in which the controlled freezing may be carried out. The ampoules, once frozen, are placed in clips which are lowered beneath the surface of the nitrogen.

To resuscitate the cells, the ampoules are removed from the liquid and placed immediately in water at 37°C. When the fluid has thawed, the cell suspension is removed from the ampoule and the cells are washed in fresh growth medium to remove the preservative. The cells may then be seeded into bottles or tubes as required.

Protective clothing, particularly a face mask to protect the eyes, must always be worn when dealing with the liquid nitrogen store. Ampoules that have not been properly sealed, and those that have developed hair-line cracks, will allow the liquid nitrogen to enter. When these damaged ampoules are removed from the store the liquid will evaporate at great speed and may cause the ampoule to explode violently. Injuries may also result from allowing the liquid nitrogen to come into direct contact with the skin. Severe burns may be sustained should this occur. Screw capped plastic ampoules are now available, these being much safer than glass.

ISOLATION AND IDENTIFICATION OF VIRUSES

It is most important that the specimens collected for virus isolation studies are taken properly and transported to the laboratory as quickly as possible. Poorly collected specimens will result in a poor isolation rate.

TRANSPORT MEDIUM

Throat swabs and other swab specimens should be collected into a transport medium. This medium consists of a balanced salt solution containing antibiotics to prevent the growth of bacteria and moulds, and a small amount (0·5%) of a protein such as bovine plasma albumen to stabilise any viruses that may be present. The end of the swab is broken off into a bottle containing 2–3 ml of this medium, and the contents of the bottle is then inoculated into cultures after the shortest possible delay. Pieces of tissue for virus studies may also be collected into this medium.

Urine for the isolation of cytomegaloviruses should be collected into a transport medium containing at least 50% sorbitol.

Urine for the isolation of viruses other than cytomegalovirus need not be collected into transport medium. Specimens of faeces are collected into waxed cartons without adding any medium. Suspensions of faeces and urine specimens should be brought to pH 7·0 prior to inoculation.

Specimens should be taken to the laboratory with the minimum delay. Specimens that are to be handled within 48 hours of collection are best stored at 4°C, and should be taken to the laboratory packed in wet ice. Should the delay exceed 48 hours the specimens should be rapidly frozen in a mixture of solid carbon dioxide and alcohol. They may then be transported to the laboratory packed in solid carbon dioxide.

Many laboratories carry out isolation studies for general practitioners. Tissue cultures may be kept at room temperature for quite long periods, up to two weeks, provided that the monolayer is adequately covered with medium. It may be possible to supply the general practitioner with a weekly supply of cultures together with instructions as to how to use them. The cultures can be inoculated at the doctor's surgery. Inoculated cultures are then returned to the laboratory for incubation.

Once a satisfactory specimen has been received in the laboratory it is necessary to decide which cultures require to be inoculated. Clear and concise clinical histories should be sent with each specimen so that some idea of the nature of the infection is conveyed to the laboratory staff.

Selected culture systems are inoculated with up to 0·2 ml of the specimen. At least two of the available cultures are inoculated and at least two tubes of each cell type are used. Uninoculated controls are incubated with each batch of inoculated tubes. It is unwise to inoculate a single cell system as there is no guarantee that a specific virus will follow the known precedents. Neither is there any guarantee that a virus indicated by the clinical diagnosis will be isolated. Inoculation of several types of cultures may, in addition, assist in the differentiation of viruses.

Inoculated cultures are observed microscopically for a minimum of 14 days, tubes being read daily. Changes occurring in the cultures are noted and the cause of any change is identified as far as possible.

Many of the common viruses will cause well-described cytopathic effects, a striking example of which is shown in Fig. 4.3. Some viruses, however, will cause such effects some time after exhibiting some other observable phenomenon such as haemadsorbtion or interference. Indeed, some viruses produce no cytopathic effect and can only be detected by indirect methods. Attempts to isolate these viruses require the use of more cultures than would normally be used, as sample cultures must be sacrificed at intervals.

Viruses of the myxovirus group, with the exception of RSV, exhibit the phenomenon of haemadsorbtion. Selected inoculated cultures and an uninoculated control are washed in cold saline and a volume of 0·5–1% guinea pig red cells is added to each tube. After 30 minutes incubation in the cold, the cultures are washed in further cold saline to remove free red cells and are examined for the presence of red cells adhering to the cell sheet. Specimens for the isolation of the myxovirus group should be inoculated into sufficient cell cultures to allow the sacrifice of one tube every three days for up to 21 days, and to leave some spare for identification of the virus should a positive be found.

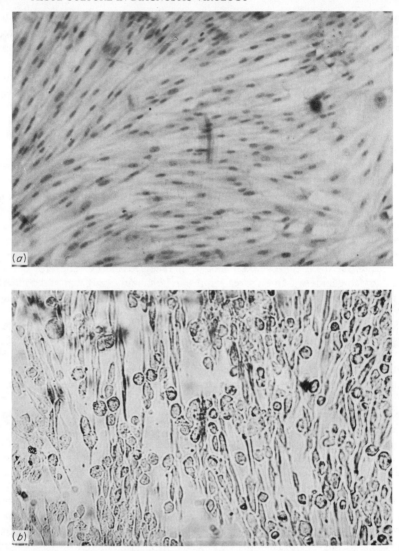

Fig. 4.3. (a) Normal human embryonic lung cells
(b) Herpes simplex virus in human embryonic lung cells.

The interference technique is the method used by one of the groups that first isolated rubella virus[13]. Tissue cultures are inoculated with specimens suspected of containing rubella virus. After a suitable incubation period, usually 10 days, a sample of the cultures is challenged with a dose of another virus that produces a rapid cytopathic effect. Interference is shown by the failure of the challenge virus to produce its usual effect in the presence of control cultures showing the expected degeneration. Challenge viruses used to detect rubella virus include ECHO 11, coxsackie A9 and Sindbis. The challenge virus need not be related to the interfering virus. Interfering agents may be identified by a modified neutralisation test.

A rapid method of detecting and typing viruses in tissue culture is by immunofluorescence. This technique is developing to a position of importance in the diagnosis of virus infection. it is not a difficult technique and requires relatively simple equipment. Viruses may be identified much more rapidly by immunofluorescence than by the normal virological methods. This technique is discussed in more detail in Chapter 8.

Tissue cultures may be used in conjunction with an electron microscope to differentiate viruses isolated from clinical material. Viruses of the Herpes group may be distinguished from pox viruses in a very short time using this instrument, on the basis of their differing morphology. It may be possible in the future to develop serological methods of typing viruses by electron microscopy, though this method has disadvantages, not least being the cost of an electron microscope.

SEROLOGY IN TISSUE CULTURES

Tissue cultures may be used in the estimation of serum antibody levels in cases where the only specimens available are acute and convalescent serum samples. Antibody to many of the enteroviruses, including the important poliovirus group, cannot be detected by any other method. The neutralisation test and its many varients consist of mainly mixing dilutions of serum with standard amounts of live virus. After an incubation period samples of the mixtures are inoculated into tissue cultures. These cultures are then incubated and are examined daily for the presence of cytopathic effect. The method is easily adaptable for use with the viruses that are detected by indirect methods. Presence of antibody is indicated by the inhibition of the effect normally produced by the virus. Varieties of the test include the haemadsorption inhibition test and plaque reduction and inhibition tests.

ANTIGEN PRODUCTION

Many laboratories will find it cheap and convenient to produce their own antigens. Reagents for use in complement fixation tests, haemagglutination inhibition tests, and neutralisation tests may be produced in the routine cultures.

Neutralisation test antigens consist of live virus of high titre. Antigens should be prepared with a titre of between 100 and 10000 × the dilution for use, though in some cases this is easier to say than to do. The greater the number of infectious particles per unit volume in these antigens, the better the results will be. The same high titre antigen can be used for much longer than one of lower titre, and better standardisation will result from its use. Live antigens are best stored at −70° C in aliquots each of which is used once and then discarded.

The complement fixing (CF) antigens of several viruses are produced in

H

routine tissue cultures. These include those of the Herpes group, the poliovirus group, respiratory syncytial virus, and the group antigen of adenovirus. Antigens to the myxovirus group are best produced in fertile eggs. Complement fixing antigen to rubella virus was first reported in 1964 by Sever *et al.*[21]. A suitable antigen may be prepared by the inoculation of a baby hampster kidney cell line BHK 21 with a high multiplicity of virus. The cells are harvested after 5–6 days, and treated with a glycine buffer at pH 9 for a period of not less than 6 hours. BHK 21 cells are a rapidly growing fibroblastic cell line. Large quantities of these cells can be produced very quickly, and they can be kept in storage until they are required. Many of the CF antigens produced in tissue cultures are best preserved by lyophilisation.

Haemagglutination inhibition antigens are not often produced in tissue cultures because higher titres of haemagglutinin are obtained using other systems. It is possible to make antigens to the myxoviruses, adenoviruses, poxviruses and the haemagglutinating members of the enteroviruses. Haemagglutinin titres of these antigens may not be high enough to make the routine use of tissue cultures for this purpose a practical procedure. Quite good titres may be found with some members of ECHO group.

The discovery of the haemagglutinin[22] of rubella virus has had a considerable effect on the laboratory diagnosis of this disease. The original method of detecting antibody to this virus was the neutralisation test, using inhibition of interference in monkey kidney cells as the indicator. This was later replaced by the method of inhibition of cytopathic effect in RK 13 continuous rabbit kidney cells[23]. Both of these methods are time consuming. The discovery, first of the complement fixing antigen, and then the haemagglutinin, made the problem of rubella antibody estimation very much easier. Haemagglutination inhibition antibodies rise very rapidly to a high titre within one week of the appearance of the rash. Thus the period of time required to establish a diagnosis of rubella has shortened from up to 14 days to a maximum of 7 days from the onset of the rash.

The antigen is produced in BHK 21 cells. A high multiplicity of virus is inoculated and the cultures are then incubated for five days. The medium is harvested and replaced with fresh fluid on days three and four, and on the fifth day the fluid and the cells may be harvested separately. The harvested fluids are treated with ether and Tween 80 and the resulting fluid is tested for its activity. The cells may be used to produce complement fixing antigen by the method of glycine extraction as described.

CONCLUSION

This chapter is an attempt to outline the uses of tissue culture in diagnostic virology. It is impossible to include all the vast amount of knowledge on this subject in one chapter, and readers are referred to the bibliography for further reading.

At the time of writing, intensive research is in progress to find effective antiviral compounds. A few agents with limited uses have already appeared. The importance of the diagnostic laboratory will increase as more of these antiviral substances are discovered. Further research into the improvement of isolation and identification techniques is required, immunofluorescent techniques and electron microscopy appear to be the most hopeful lines at the present time. Until some alternative method of isolating and identifying viruses is discovered, tissue cultures will continue to form the basis of all work in this field.

REFERENCES

1. ENDERS. J. F.. WELLER. T. H. and ROBBINS. F. C.. *Science, N.Y.*, **109**, 85 (1949)
2. DALLDORF, G., *Science, N.Y.*, **110**, 594 (1949)
3. MELNICK, J. L., SHAW, E. W. and CURNEN, E. C., *Proc. Soc. exp. Biol. Med.*, **71**, 344 (1949)
4. ROBBINS, F. C., ENDERS, J. F., WELLER, T. H. and FLORENTINO, G. L., *Am. J. Hyg.*, **54**, 286 (1951)
5. ROWE, W. P., HUEBNER, R. J., GILMORE, L. K., PARROT, R. H. and WARD, T. G., *Proc. Soc. exp. Biol. Med.*. **84**, 570 (1953)
6. HILLEMAN, M. R. and WERNER, J. H., *Proc. Soc. exp. Biol. Med.*, **85**, 183 (1954)
7. CHANOCK. R. M.. ROIZMAN. B. and MYERS. R.. *Am. J. Hyg.*, **66**, 281 (1957)
8. SMITH, M. G., *Proc. Soc. exp. Biol. Med.*, **92**, 424 (1956)
9. BICKERSTAFF, E. R., *J. Path. Bact.*, **66**, 391 (1953)
10. MOGABGAB, W. J., GREEN, I. J., DIERKHISING, O. C. and PHILLIPS, I. A., *Proc. Soc. exp. Biol. Med.*, **89**, 654 (1955)
11. MOGABGAB, W. J., SIMPSON, G. I. and GREEN, I. J., *J. Immun.*, **76**, 314 (1956)
12. SHELEKOV, A., VOGEL, J. E. and CHI, L., *Proc. Soc. exp. Biol. Med.*, **97**, 802 (1958)
13. PARKMAN, P. D., BUESCHER, E. L. and ARTENSTEIN, M. S., *Proc. Soc. exp. Biol. Med.*, **111**, 225 (1962)
14. HOORN, B. and TYRRELL, D. A. J., *Arch. Virusforsch.*, **18**, 210 (1966)
15. MORGAN, J. F., MORTON, H. J. and PARKER, R. C., *Proc. Soc. exp. Biol. Med.*, **73**, 1 (1950)
16. EAGLE, H., *Science, N.Y.*, **130**, 432 (1959)
17. SMITH, C. E. G.. SIMPSON, D. I. H., BOWEN, E. T. W. and ZLOTNIK, I., *Lancet*, **ii**, 1119 (1967)
18. GEY, G. O., COFFMAN, W. O. and KUBICEK, M. T., *Cancer Res.*, **12**, 264 (1952)
19. FJELDE, A., *Cancer, N.Y.*, **8**, 845 (1955)
20. EAGLE, H., *Proc. Soc. exp. Biol. Med.*, **89**, 362 (1955)
21. SEVER, J. L., HUEBNER, R. J., CASTELLANO, G. A., SARMA, P. S., FABIYI, A., SCHIFF, G. M. and CUSUMANO. C. L.. *Science, N.Y.*, **148**, 385 (1965)
22. STEWART, G. L., PARKMAN, P. D., HOPPS, H. E., DOUGLAS, R. O., HAMILTON, J. P. and MEYER, H. M., *New Engl. J. Med.*, **276**, 554 (1967)
23. MCCARTHY, K., TAYLOR-ROBINSON, C. H. and PILLINGER. S. E., *Lancet*, **ii**, 593 (1963)

BIBLIOGRAPHY

HORSFALL, F. L. and TAMM, I. (Eds.) *Viral and Rickettsial Infections of Man*, 4th edn, Pitman Medical, London (1966)
ANDREWES, C. E. and PERIERA, H. G., *Viruses of Vertebrates*. 2nd edn, Ballière, Tindall and Cassell, London (1967)
HOSKINS. J. M.. *Virological Procedures*, Butterworths, London (1967)
PAUL, J.. *Cell and Tissue Culture*, 4th edn, Livingstone, Edinburgh (1970)

FIVE

The cultivation of mammalian macrophages *in vitro*

GERALD D. WASLEY and ROLAND JOHN
St. Thomas's Hospital Medical School, London

INTRODUCTION

In this chapter a few well-tried methods for the artificial cultivation of mammalian macrophages will be described in such detail as to ensure the provision of a sound technical basis for anyone who may wish to take up the study of these cells by this means. It is important to be clear at the outset that the term cultivation is, in this particular connection, restricted in connotation to mean no more than keeping the cells alive and healthy by artificial means, for it is generally true that, at any rate under *in vitro* conditions, normal fully different-iated mammalian macrophages are incapable of multiplication.

THE GENERAL CHARACTERISTICS OF MACROPHAGES

Before undertaking the description of the technical procedures entailed in the *in vitro* cultivation of mammalian macrophages, it may not come amiss to give a brief account of the characteristics of these quite remarkable cells. They are widely distributed in the bodily tissues. They have the power of amoeboid movement—a function, however, by no means at all times in evidence. They are possessed of the ability to ingest and subsequently, by means of intracellular enzymes, often to digest both particulate and dissolved matter whether of autochtonous or foreign origin. Of mesenchymal derivation, they are of the lymphoid series of cells and, by virtue of their phagocytosis of matter that may be antigenic, play an essential part in the production of antibodies together with the lymphocytes and plasma cells.

The name macrophage was invented by Elie Metschnikoff[1] who was the first to appreciate the importance of phagocytosis in the defence of multicellular animals against invading pathogenic micro-organisms. He used the term to distinguish these cells in the higher animals from the amoeboid phagocytic polymorphonuclear leucocytes of the blood which, being of smaller size, he called microphages.

Cells resembling the macrophages of the higher animals are to be found throughout the animal kingdom, in even the simplest of organised multicellular animals. In these latter lowly creatures, which have either no circulatory system or one in which the circulating liquid is devoid of cells of microphage type, they inhabit the connective tissues. They wander freely through the interstices concerning themselves with the disposal of injured or effete tissue elements and any foreign matter, including micro-organismal parasites, which may have gained entry into the tissues. In the exercise of these functions, they are guided chemiotactically to the objects of their attention and often, especially in microbic infections, evince a facility for rapid multiplication.

In the course of evolution, these amoeboid phagocytes have persisted and retained most of their ancient functions. However, with the progressive elaboration of the various organs, some of them have come to be incorporated in the structure of certain of the latter where they exert their phagocytic power without, other than in exceptional conditions, moving away from their appointed places—these are the so-called fixed macrophages of the blood and lymph sinusoids.

THE MACROPHAGES OF THE CONNECTIVE TISSUES

In the higher animals including man macrophages are present just as in the simpler creatures, in the ordinary loose connective tissues in which they are distributed widely, though somewhat irregularly, and in considerable numbers. They are particularly numerous in the adventitial tissue of blood vessels, where many of their precursor cells are also located, and especially so in certain anatomical sites such as the subserous tissue of the peritoneum and pleura.

In normal conditions the connective tissue macrophages are inactive and are known as resting wandering cells or adventitial cells. In this dormant state their shape is largely dependent upon the extent to which the tissue in which they lie impinges upon them. In the denser tissues they are flattened and their cytoplasm is extended into several usually short and blunt but sometimes fine long-drawn-out and branched projections, whereas in looser tissues they tend to be more rounded with fewer cytoplasmic projections. When suitably stimulated, as happens, e.g. at some stage in most inflammatory reactions, the macrophages of the affected part retract their cytoplasmic projections, become actively amoeboid and phagocytic and may increase in numbers by, at least in part, the multiplication and differentiation of the neighbouring precursor cells. In this active condition, when they are commonly referred to as histiocytes[2], they sometimes accumulate in such numbers and lie in such close proximity to one another as to present an appearance resembling that of an epithelium, and then they are called epithelioid cells. Also in certain inflammations, e.g. tuberculosis, and when dealing with large fragments of foreign matter several of them may fuse together to produce a multinucleated or giant cell. Ranvier[3] noticed that active macrophages sometimes separate off fragments of their cytoplasm and on this account named them clasmatocytes—a term now seldom, if ever, used. This process, called clasmatosis, has been observed in rabbit-ear-chamber preparations and also in artificial culture though its significance remains uncertain.

The ordinary loose connective tissue is not, in the higher animals, the only source of these mobile macrophages—neither is it the only site in which they function. Some arise from precursor cells in the haemopoietic marrow, spleen and probably the lymph nodes and enter the blood stream, where they are known as monocytes, or large mononuclear leucocytes, and constitute in man about 5% of the white blood cells. In many inflammatory reactions, these circulating macrophages migrate through the walls of the small vessels of the affected part to enter the connective tissue spaces. There they augment the numbers of the already present and activated local and locally produced macrophages from which they become indistinguishable in both form and function.

Macrophages are widely distributed in the tissue of the central nervous system which they enter during the course of embryonic development[4] and in which, in the resting condition, they are known as microglial cells. Macrophages and their precursors are also present in the pulmonary connective tissue. From there they regularly migrate into the alveoli to perform the essential task of ridding the latter of any finely particulate matter, including bacteria, that may be inhaled and deposited upon the alveolar surfaces. These cells also phagocytose various endogenous substances that may escape into the alveoli in disease, such as red cells in cases of congestive cardiac failure. Known as alveolar phagocytes they are also called, according to the circumstances prevailing, dust cells and heart-failure cells. In somewhat similar ways other surfaces liable to contamination, such as those of the pharynx and intestinal tract, gain a measure of protection from the wandering macrophages of the subepithelial connective tissues.

THE MACROPHAGES OF THE BLOOD AND LYMPH SINUSOIDS

Whereas amoeboid movement is an essential part of macrophage function in those tissues so far considered—the cells migrate to the site where their phagocytic action is needed—the converse holds in the case of the fixed macrophages for they are normally stationary and the matter destined to be treated by them is brought to them by the blood and lymph streams. These latter cells, as has already been briefly mentioned, are incorporated in the structure of specialised blood and lymph channels known as sinusoids. The blood sinusoids are defined vessels which are lined not with ordinary vascular endothelium but with fixed macrophages and form the intimate vasculature of a certain few tissues: the haemopoietic marrow, the red pulp of the spleen, the parenchyma of the liver (here they are named after von Kupffer[5] who first observed them), the adrenal cortex and medulla, the anterior lobe of the pituitary, the parathyroid glands and the coccygeal body. These specialised vessels are, therefore, the counterparts of the capillary blood vessels of the other tissues of the body. The lymph sinusoids are the pathways along which passes the greater part of the lymph that traverses the lymph nodes. They are little more than tracts in the lymph-node tissue where the reticulin framework of the latter is more coarsely meshed than elsewhere and to the fibres of which the macrophages are attached.

THE MACROPHAGE SYSTEM AND ITS FUNCTIONS

It has long been accepted, largely as a result of the use of vital-staining methods, that the macrophages, whether wandering or fixed, are all fundamentally similar. The work that led to this conclusion was reviewed by Aschoff[6] in 1924. Since that time much has been done towards clarifying the generative relationships between the macrophage and the other cells of the haemopoietic and lymphoid tissues, and interest has turned from these more micro-anatomical matters to the functions of the macrophage system. Its role in infection including the production of immune bodies, in the metabolism of the blood pigments, in disorders of lipoid metabolism and in the metabolism of chemotherapeutic agents, as well as the biochemical and biophysical qualities and the hyperplastic and neoplastic potentialities of its cells, have all been intensively studied. Much of this later work is well covered by the reviews of Cappell [7, 8], Downey[9] and Jacoby[10] and by the monograph by Nelson[11].

THE MACROPHAGE IN ARTIFICIAL CULTURE

The method of artificial cultivation has played and continues to play an important part in the investigation of macrophage behaviour and function. In the early applications of this method explants of material containing macrophages were cultivated in natural culture media of which plasma was commonly used. When cultures are set up in this way, the plasma promptly clots and the cells are to be observed amidst a tangle of fibrin filaments which more or less obscure the view of them and, by impinging upon them in much the same way as do the surrounding tissue elements upon the resting wandering cells *in vivo*, cause their morphology to be very variable. Some of the cells eventually migrate out of the clot on to the adjacent clear glass of the culture-vessel wall and this greatly simplifies microscopical observation of them, for not only are they unobscured by the fibrin but also they all lie in substantially the same plane. Further, their shapes are uninfluenced by the very variable distorting effects of the fibrin meshwork. This advantageous outcome accrues, however, at the expense of the time taken for the emigration to occur and, in the case of mammalian macrophages, this is a limiting factor, for the inability of these cells to multiply under artificial conditions determines the length of life of any culture. The difficulty is overcome by using the monolayer method of cultivation in which the cells are suspended in a medium that remains liquid and in such numbers that when they settle on the floor of the culture vessel, they do so in a single layer and without overcrowding. This method greatly facilitates the study of individual cells and the interactions between the cells in both normal and experimentally produced abnormal cultural conditions.

Whatever its source tissue, the living macrophage is, when inactive and suspended in a physiologically constituted liquid medium, practically spherical—just as it is when in monocyte form in the circulating blood—and of a diameter varying between 15 and 25 μm. By means of the phase-contrast microscope, the nucleus is seen to contain a single nucleolus, to be either oval or indented (reniform) and to be possessed of a rather open-textured chromatin network which may, however, be more compact while the nucleus is correspondingly smaller and usually more rounded. The nucleus is eccentrically placed in the cell cytoplasm and, if indented, is disposed with its indentation facing towards the farther side of the cell. The cytoplasm may contain small vacuoles and amorphous granules and the cytocentrum, which is a remarkably large body that makes incessant to-and-fro movements, is situated near the nucleus where, if the latter be indented, it lies in the region of the cytoplasm adjacent to the indentation.

In suitably fixed and stained preparations the cytoplasm is somewhat basophilic and the cytocentrum is seen to contain two distinct centrioles and to be surrounded by the Golgi apparatus and most of the mitochondria.* Electron microscopy reveals the presence in the cytoplasm of numerous lysosomes which have been found to contain various hydrolysing enzymes such as acid phosphatase, cathepsin, deoxyribonuclease, ribonuclease and various esterases as well as lipase.

When living macrophages, suspended in a suitable liquid medium, are allowed to deposit at body temperature on a surface such as the glass of a culture vessel, they gradually lose their globular form, become more or less flattened, throw out numerous short pseudopodia, thus resembling resting wandering cells and, in the course of a total time of 2–3 hours, adhere to the surface. The strength of this adhesion is remarkable—it is greater, probably, than that achieved by any other type of cell similarly circumstanced—and is of such a degree that the surface may be rinsed with medium without fear of dislodging the cells. The whole of this process occurs most rapidly at body temperature, is inhibited by cold and takes about six times as long if the surface concerned be, like that of most plastics, not water wettable.

After settling in this way the cells display a striking feature which is peculiar to macrophages—the 'membrane ondulante' of Carrel[15]. This structure, which is readily seen by means of the phase-contrast microscope, is the narrow, completely transparent and greatly thinned outer edge of the cytoplasm of the flattened cell which is expanded into ruffle-like folds that are in continuous undulating movement. While the cell is stationary, the numerous pseudopodia referred to above are seen to be continually thrown out, retracted and reformed at ever-changing sites along the cell margin and each of these transient

---

* The macrophages of the cat, man, the mouse, the rabbit, the rat and, presumably, those of other mammals can be readily distinguished from all other cells, with the sole exception of the oligodendroglia by means of the silver-impregnation methods of del Rió Hortega and de Asúa[12] and of Weil and Davenport[13]. These methods combined with the use of the flying cover-slip technique (see page 112) can be of the greatest value for recognising macrophages with certainty in cultures containing other varieties of cell—see Marshall[14] for references to the literature and a description of the methods with an account of some useful modifications of them.

projections is tipped with an undulating membrane. When in amoeboid motion a solitary large pseudopodium is extended in the direction of the motion, the whole cell becomes elongated in consequence and the undulating membrane is confined to the advancing front of the pseudopodium.

In monolayer cultures the cells may present a variety of forms which seem to depend to some extent on the closeness of the cells to one another and on the age of the culture. In young cultures and in the absence of overcrowding and any apparent stimulus to migration, the cells may remain stationary or move about in a random fashion. When stationary they are liable to extend several pseudopodia some of which may become very long and fine and all of which are tipped with an undulating membrane. Occasionally one of these longer pseudopodia may undergo clasmatosis. When the cells are so crowded that they are in contact with one another they put out numerous short pseudopodia and become immobile though the ruffled edges of the pseudopodia continue in constant motion. In this epithelioid condition they tend to adhere to one another and may do this so firmly that they can be stripped off the culture-vessel surface as an unbroken membrane. This mutual adherence is probably brought about by entanglements between the pseudopodia of contiguous cells and has its counterpart *in vivo* in the macrophage lining of the blood sinusoids. In overcrowded older cultures some of the epithelioid cells may fuse together to form giant cells.

When the amoeboid motion of a macrophage is not random but directed to the phagocytosis of some particulate object, one major pseudopodium is extended as described above, but, in addition, the whole of the rest of the cell membrane can be seen to be throwing out numerous minute continuously fluctuating projections. Having been reached, the particle becomes trapped between some two of these minor pseudopodia and then appears to sink into the cell cytoplasm where it comes to lie within a vacuole in which, if it be so susceptible, digestion of it occurs. If the particle proves to be indigestible it is likely to be retained in the cell indefinitely. Should much matter be engulfed, then it or the products of its digestion accumulate within the cell which may become greatly enlarged in consequence. The mechanisms of phagocytosis and digestion by the macrophage have been elucidated by means of electron microscopy: a description of them is to be found on cell structure such as that by Toner and Carr[16].

SOURCES OF SUPPLY OF MACROPHAGES FOR CULTIVATION

From what has been said already it will be clear that most animal tissues could be used as sources of macrophages. These cells were, in fact, recognised during the early days of the practice of tissue culture in explants of many different tissues taken from a wide variety of animals. They were first grown in pure culture in 1921 by Carrel and Eberling[17] from an explant of the buffy coat of the centrifuged blood of an adult domestic fowl using homologous plasma containing embryonic tissue juice as the culture medium. This achievement was the result not only of patient experiment but also of the essential fact that the large mononuclear leucocytes (blood macrophages) of avian blood are capable of multiplication as well as continued survival in artificial media, whereas the other types of avian leucocyte are not. Thus, by repeated subculture, the macrophages were obtained unmixed with other cells. Carrel and Eberling succeeded in maintaining these artificially cultured macrophages in a healthy condition for as long as 3 months and so, in a series of experiments, were enabled to study their morphology and amoeboid and phagocytic activities as well as their reactions to variations in the composition of the culture medium.

Since much of the work done with macrophage cultures has some bearing on human physiological and pathological problems, it is appropriate that the cells used for such work should be human or at least from some other mammal. The attempt to fulfil this requirement is met at once by the practical difficulty that, although they will survive in a healthy condition in artificial media, it has so far not been found possible—there are one or two doubtful exceptions—to get macrophages to multiply in such circumstances. This failure is probably not a technical one but merely a reflexion of the likelihood that, even in the normal course *in vivo*, the mammalian macrophage attains full differentiation only at the expense of losing its power of reproduction. Whatever the truth of the matter may be, this difficulty has led to a search for sources that would yield macrophages in numbers adequate for the intended use of the cells when cultured but accompanied by minimum numbers of cells of other types and would, at the same time, be readily accessible without undue risk of micro-organismal contamination. As a result it has been found that of the mammals conveniently available those most suitable are, in descending order of suitability and, therefore, of frequency of use, the mouse, the rabbit, man, the guinea-pig and the rat. Each of the many tissues of these animals from which the cells might be obtained manifests some practical disadvantage. The yield of cells may be very small—a serious shortcoming since the cells do not multiply in culture. The material collected may inevitably be contaminated with other kinds of cell—according to the nature of the experiment to be made with the established culture this may not be a serious disadvantage but if it be so, then some procedure, which may be fairly complex, for separating the macrophages from the other cells must be undertaken. The source tissue may be so situated anatomically that the method used to obtain from it material free of micro-organismal contamination has of necessity to be elaborate—a wearisome complication if repeated recourse to the tissue be needed as when a succession of cultures has to be made.

By way of illustration of the foregoing, there follows a brief account of the pros and cons of a small selection of sources in common use, including those which provide an acceptable compromise and are described in detail in the section on practical procedure.

Whole blood is an obvious source and is easily collected aseptically but the procedure for the separation of the macrophages (monocytes) from the other cells is complex and the yield is low. Haemopoietic marrow contains many macrophages but, as with solid-tissue sources in general monolayer cultures of these cells can only be obtained as a result of their emigration from an explant of the tissue and are invariably contaminated with fibroblasts which multiply in the culture and tend to swamp the macrophages.

The peritoneal transudates of most mammals contain macrophages and those of the mouse and guinea-pig are normally quite rich in these cells though they also contain a larger number of a mixture of macrophage precursors, lymphocytes, eosinophil polymorphonuclear leucocytes and mast cells. Since these intraperitoneal cells can be washed out of the peritoneal cavity aseptically and without much difficulty and the contaminating cells eliminated quite easily, this is a very commonly used source. Mice are much cheaper than guinea-pigs and as, for this purpose, the latter offer no advantages over the former the mouse is the animal generally used.

It may be as well to refer at this juncture to the practice of some workers of bringing about an increase in the macrophage content of the peritoneal liquid by means of the intraperitoneal injection of a foreign substance some time, usually a few days, before the peritoneal cavity is to be washed out. A variety of substances has been used: in the mouse, starch[18], sodium caseinate[19] and Difco thioglycollate medium[20]; in the guinea-pig, glycogen[21]. Similar procedures have been used with other animals, for example: the rabbit—glycogen[22], light mineral oil[23], heavy mineral oil[24] and liquid paraffin[25]; the rat—rabbit-liver glycogen[26], beef-heart-infusion broth fortified with Difco proteose peptone[23] and proteose peptone[27]. These treatments certainly increase the yield of macrophages especially in animals whose peritoneal transudate is not normally rich in these cells. It is, however, open to question whether macrophages that have reacted in this way will react entirely normally to other agents that may be applied to them experimentally in subsequent culture. Further, if the method to be described for obtaining peritoneal macrophages from the mouse be used, it will be found that the yield of cells is adequate without resort to any ancillary treatment.

The liquid withdrawn from the peritoneal cavities of human patients undergoing the peritoneal-dialysis treatment of chronic renal failure, is a very valuable source for it often contains many macrophages and is usually pretty free of other cells. Its use is necessarily limited to laboratories situated close to the places where the treatment is carried out but this restriction is a diminishing one for such centres are being set up in increasing numbers and are now established in most teaching as well as many other large hospitals. It is likely, therefore, that increasing use will be made of this material so that a warning of its sometimes dangerous nature will not be out of place. In cases of chronic renal failure, the immunity mechanisms tend to be depressed and because of this, so it is thought, the patients are rendered unduly liable to become persistent carriers of the causative agent of homologous-serum jaundice whether they have suffered overtly from the disease or not. The agent is present in the blood and so contaminates the apparatus used in the treatment and, since it is very highly infective and unusually thermostable, it is liable to be transferred in active form not only to other patients but also to all who may come into contact with the blood, and, quite possibly, the peritoneal dialysate, and to those who are responsible for the maintenance of the apparatus. Since this disease can be very severe sometimes causing death in the acute phase or leading to chronic, progressive and ultimately fatal liver damage, these peritoneal dialysates must be handled with the greatest care (see page 133 for a description of the precautions to be taken).

The lung normally contains many free macrophages—the alveolar phagocytes—and Myrvik, Leake and Fariss[28] have devised a comparatively simple technique for obtaining them by washing out the alveoli and airways of rabbit lungs. We have not tried the method ourselves but it would appear that this source, despite certain disadvantages which will be referred to later, is a very good one for it gives a high yield of macrophages and the contaminating cells are not only small in numbers but, being almost wholly neutrophil polymorphonuclear leucocytes, are easily disposed of.

Detailed descriptions of the practical procedures used for obtaining macrophages from the last three of the above-mentioned sources, namely, the mouse and human peritoneal cavities and the rabbit lung, and for setting up the cultures of the cells are given on pages 128–136.

APPARATUS AND MATERIALS

GENERAL

Everything that may come into contact either directly or indirectly with either the cells or the media in which the latter are to be cultivated must be prepared to those high standards of cleanliness and sterility which are indispensable in all kinds of tissue culture work.

Glassware should be washed with a free-rinsing detergent such as Pyroneg* and very thoroughly and repeatedly rinsed with double-glass-distilled or di-ionised water. When protein or other matter has been allowed to dry out on glass surfaces on which the cells are to be cultivated vigorous brushing may be needed to remove it. Every care must be taken to remove such matter completely otherwise streaks of it left by the brush bristles (brush marks) form pathways along which the cells tend to migrate thus producing an unnatural pattern of cell distribution which may lead to the formation of erroneous conclusions (Fig. 5.1). The use of such cleaning agents as chromic acid is to be avoided on account of the difficulty of rinsing them away completely. All glassware, such as that of culture tubes, which is subject to repeated washing with detergents should be preferably of borosilicate glass which is not liable to corrosion by the phosphates present in most detergents.

Effects similar to those caused by brush marks are produced by surface scratches (Fig. 5.2) and, additionally in the case of plastics, mould marks. The relevant surfaces must, therefore, be scrutinised and be seen to be free of such imperfections.

Bungs and other flexible articles such as tubing and the cap liners of screw-top bottles are best made of silicone rubber. Non-toxic white rubber is also suitable but, though initially very much cheaper than silicone rubber, is far less durable under conditions of oft-repeated heat sterilisation.

The sterilisation of glassware is better carried out by dry heat—180°C for $1\frac{1}{2}$ hours in the hot-air oven—than in the autoclave. The reason for this preference is that the water used in autoclaves commonly contains some amount of foreign matter which is liable to be carried up in the steam and be deposited on the surfaces of the articles being sterilised. Also during drying off, such deposits

---

* Pyroneg is made by and may be bought from Diversey Ltd., Star House, Mutton Lane, Potters Bar, Hertfordshire.

tend to run into streaks which, present on surfaces on which cells are to be cultured, may lead to artefacts similar to those caused by brush marks and scratches.

Silicone rubber withstands repeated hot-air sterilisation at the temperature and for the duration specified above without deterioration but non-toxic white rubber does not. Sterilisation of the latter material must be accomplished,

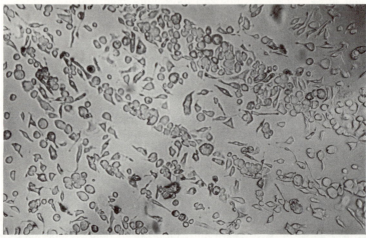

Fig. 5.1. Effect on distribution of cells of presence of streaks of residual foreign matter (brush marks) on culture-vessel surface. Transmural photomicrograph of live simple-tube culture of mouse-peritoneal macrophages. Objective × 10, Ocular ×8

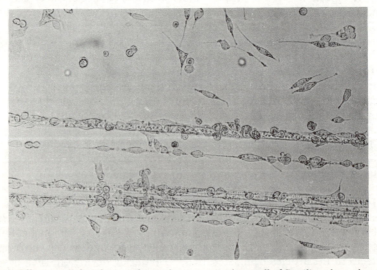

Fig. 5.2. Effect on cell distribution of scratches on cover-glass wall of Sterilin culture chamber. Photomicrograph of living mouse peritoneal macrophages. Objective × 10, Ocular × 8.
The cover-glass in question was not examined with sufficient care before use.
Note improved quality of image compared with Fig. 5.1

therefore, by other means: autoclaving at $0 \cdot 7$ kg cm^{-2} for 10 minutes is satis-factory and does not occasion too rapid deterioration.

Some varieties of the plastic culture chambers to be described later are sterilised by gamma irradiation during the course of their manufacture and are ready for use as supplied. They are to be used once only. Others not so treated can be sterilised by immersing in 70% absolute alcohol in double-glass-distilled or de-ionised water for 30 minutes followed in turn by rinsing 4 times with similarly purified water, drying in air at room temperature in a sterile chamber and finally irradiating with ultraviolet light for 30 minutes by means of a Phillips germicidal lamp.

CULTURE VESSELS

Several types ranging between simple round-bottomed glass tubes and more elaborate built-up vessels known as culture chambers are used.

CULTURE TUBES

The simple tube is satisfactory for most types of work. It is cheap, easily manipulated and readily cleaned and re-sterilised after use.

The tubes, preferably of boro-silicate glass, should be 100 mm long and of 12 mm diameter and fitted with rubber bungs. If the bungs be of silicone rubber then each tube and its bung, the latter loosely fitted, can be sterilised together in the hot-air oven.

In use $1 \cdot 0$ ml of culture medium (see p. 128) containing the cells to be cultivated in appropriate concentration (see p. 131) is placed in each tube, the atmospheric air in the latter displaced with sterile air containing about 5% of carbon dioxide—a procedure known as 'gassing',* the tube stoppered and the batch of tubes supported for incubation in a rack designed to secure an angle of 3° between the long axes of the tubes and the horizontal. The uppermost aspect of each tube as it lies in the rack is marked near the bung with a grease pencil. The cells present deposit on and, after incubation for 2–3 hours, the macro-phages among them adhere to the lower surface of the tube to form a monolayer so that, at any required later time, simply by rotating the tube through 180° or thereabouts the macrophages may be observed in the living condition through the tube wall by means of an ordinary or phase-contrast microscope equipped with a substage condenser of long working distance. The examination having been completed the tube is replaced in the rack with the grease-pencil mark upwards thus ensuring the re-immersion of the cells in the medium.

It will be realised, of course, that in using this method of examination the presence of the intervening tube walls, because of the latters' curvature and substantial as well as unavoidably irregular thickness, necessarily diminishes the quality of the microscopical image. Also the thickness of the tube wall, by

---

* Gassing has to be repeated from time to time during the course of cultivation. It is called for whenever the pH of the medium, signalled by an indicator included in the latter, begins to fall. See also page 126.

encroaching on the free working distance of the microscope objective, precludes the use of objectives of magnifying powers in excess of about 10 diameters.

Despite these disadvantages observations made in this simple way can be very useful, for in many experiments examination of no more than the grosser features of the cultured cells is needed and for this purpose the method is wholly adequate. When more critical microscopy or high-quality photomicrography of the living cells is required, resort must be made to the use of some form of culture chamber whose transparent walls, to the inner surface of one of which the cells are arranged to adhere, are thin, optically flat and parallel to one another (compare quality of cell images in Figs. 5.1 and 5.2).

A very useful addition to the simple tube consists in the placing in the tube, with one end resting on the bottom of the latter, of an oblong piece of coverglass of such a width as to allow of its immersion in the medium and of its easy insertion and withdrawal. When the inoculated medium is run into a tube so equipped some of the macrophages in it deposit on and adhere to the upper surface of this 'flying' cover-slip which may be removed at any time during the life of the culture, the adherent cells appropriately fixed and stained and the cover-slip mounted on a slide to make a preparation suitable for critical microscopical examination. Meanwhile those of the cells introduced that fell to the lower wall of the tube remain undisturbed so that their cultivation may be continued if desired.

The Leighton modification of the simple tube is preferred by some workers and consists of an oblong flat-surfaced extrusion of one side of the lower end of the tube (Fig. 5.3)*. This tube, which is available in various sizes and either

Fig. 5.3. The Leighton-type culture tube

plain mouthed or screw capped is incubated in the horizontal position with the extrusion or 'window' downwards. The medium is wholly contained within the extrusion the cells depositing on and adhering to its inner flat surface. The microscopical images of the cells when viewed through the window are, compared with those formed through the wall of a plain tube, improved to the extent that the element of distortion caused by the curvature of the latter is eliminated. This type of tube is somewhat more difficult to clean thoroughly and appreciably more expensive than the simple tube.

CULTURE CHAMBERS

The essential optical feature of a culture chamber intended for use in the study

* Leighton tubes are made by and may be bought from Johnsen & Jorgensen Ltd., Herringham Road, London, S.E.7.

of macrophages cultivated in a liquid medium is its possession of at least one wall—the lower, to which the cells having deposited adhere—that is for all practical purposes optically flat and of a transparent material of such refractive index and thickness that, with the chamber inverted for the microscopical examination of the cells, its interposition between the microscope objective and the cells has no deleterious effect on the image formed of the latter. In all of the several varieties of chamber that have been devised from time to time this essential requirement is met by using a cover-glass to form one wall. Obviously it is also very desirable that the opposite wall—the lower when the chamber is inverted for microscopy—be of a similar physical nature to and be arranged parallel to the cell-bearing wall and close to it if a substage condenser of normal working distance is to be used and correctly focussed. Very few chambers fulfil all three of these latter desiderata and in those potentially capable of doing so the dimensions of the structure in which the actual chamber is mounted are, for simple mechanical reasons, almost unavoidably such that the lower wall of the chamber proper lies some distance above the level of the under surface of the mount. It follows, therefore, that, unless the range of vertical adjustment of the substage condenser be such that the latter can be projected the required amount above stage level, the accurate focussing of an ordinary condenser is impossible and the use of a special condenser with a long working distance must be resorted to *. Even when a condenser of long working distance is used it is optically more satisfactory to have the upper and lower chamber walls pretty close together and in the better designs they are separated by no more than a millimeter or so.

In addition to the enhancement of their optical qualities other improvements have been effected in the design of certain chambers which greatly facilitate their use for the maintenance of cultures over periods as prolonged as 2 to 3 weeks. In these further improved forms an additional space, so arranged as to surround the margin of the shallow space between whose walls the medium and cells are confined, provides accommodation for an atmosphere which, being in contact with the 'edges' of the medium, permits of gaseous interchanges between it and the medium. This gas space, as it is called, is furnished with some form of communication with the exterior of the chamber so devised that, by its means and using ordinary care, the composition of the atmosphere can be controlled without risk of exposing the culture to infection. The same or some additional external communication provides for the passage into the chamber of hypodermic needles through which the initial introduction of medium and cells is conveniently made and, subsequently and without risk of infecting or disturbing the cells, exhausted medium withdrawn and replaced with fresh.

* In this connection it may be worth recalling that the required extra range of upward condenser adjustment was present on all those prewar Watson microscope stands which were supplied equipped with that maker's 'Service' pattern attachable mechanical stage. This stage was constructed on the lines of the 'in-built' type of mechanical stage and was used mounted on top of the standard plain stage. Its working surface was about 5·0 mm above that of the plain stage and the range of upward condenser adjustment relative to the latter was increased accordingly. Thus, if one of these stands be available, and many of them are still in use, the removal of the mechanical stage and the mounting of the culture chamber on the plain stage will probably permit of the correct focussing of a condenser of normal working distance.

Control of the atmosphere and renewal of the medium may be carried out intermittently or, in the case of the more elaborate chambers, continuously.

These better-designed chambers, by providing ready access to gas space and culture, facilitate the experimented application to the latter of abnormal substances, gaseous or liquid, in solution or suspension, while their good optical and other qualities make possible accurate and prolonged microscopical observations of the results of the action of such substances on the living cells. Two such chambers will now be described, both in their original form and as modified for commercial manufacture.

THE TREVAN AND ROBERTS AND PRIOR CULTURE CHAMBERS

The chamber devised by Trevan and Roberts[29]* is very satisfactory for, though somewhat complex and a little delicate in construction and, in the absence of the extra degree of condenser adjustment referred to above, needing a condenser of long working distance (the under surface of the lower wall of the chamber is 4·4 mm† above the level of the under surface of the chamber-housing mounting plate), it is still fairly simple to use and provides the means for carrying out a wide range of experimental work. Originally designed for use with an inverted microscope it works, nevertheless, perfectly well, the condenser difficulty apart, with a conventional erect instrument. Used in this way, however, the chamber itself must be inverted during incubation, at least for the first few hours, for then the macrophages deposit on and adhere to the inverted upper wall where, on turning the chamber the right way up, they are suitably placed for examination with the erect microscope.

The chamber is designed to be used repeatedly and can be completely dismantled for cleaning. It comprises two 32 mm diameter No. 1 circular microscope cover-glasses separated about 2 mm‡ from one another, to form the upper and lower walls, by a peripherally placed silicone-rubber sealing ring of circular cross section and supported, with moderate compression of the ring, between the clamped-together upper and lower parts of an annular aluminium-alloy housing of just under 40 mm external diameter. The two parts of the housing are held together by three fine screws (Fig. 5.4). In preparation for use the peripheral part of the internal aspect of each cover-glass is lightly smeared with silicone stop-cock grease leaving clear an approximately circular central area about 10 mm diameter. By this means the medium and cells, when introduced, are confined to the central part of the chamber while leaving unoccupied the remaining surrounding part which serves as the gas space.

The lower which is the deeper part of the housing is provided with two radially disposed stainless-steel side tubes furnished with externally knurled, taper-fitting aluminium-alloy§ caps and arranged, in the original design, 60°

* The Trevan and Roberts chamber is manufactured in slightly modified form by and may be bought from W. R. Prior & Co. Ltd., London Road, Bishop's Stortford, Hertfordshire by whom it is called the Prior Tissue Culture Chamber.

† 4·9 mm in the Prior modification.

‡ About 1·9 mm in the original design and about 2·2 mm in the Prior chamber.

§ Brass in the Prior chamber.

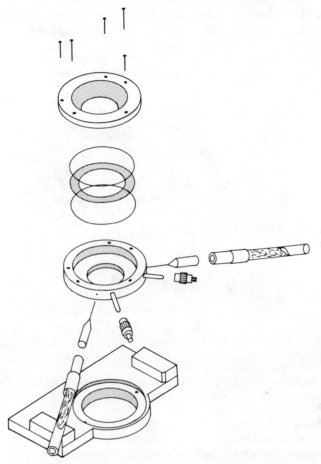

Fig. 5.4. Exploded drawing of Trevan and Roberts culture chamber including gas needles and filters.

apart but in the Prior modification diametrically opposite each other (Fig. 5.5)—a rearrangement that leads to a certain difficulty which will be referred to later. It is better that the turning motion applied to the caps, either to secure them firmly on or to release them from the tapered ends of the tubes, be always in the clockwise direction in order to avoid the risk of loosening the tubes which are screwed into the housing. Each cap (in the Prior modification) is provided at its outer closed end with a pair, one on either side, of parallel flat surfaces by means of which it can be gripped firmly with forceps. The bore of the tubes is only fractionally less than the depth of the chamber and each tube is so placed that the vertical diameter of its internal orifice almost exactly coincides with the gap between the two cover-glasses so that the compressed sealing ring abuts nicely against the orifice. Access to the central part of the chamber for the purposes of introducing the medium and cells, renewing the medium etc. is gained by passing a 38 mm ($1\frac{1}{2}$ in) long, 0·55 mm (24 S.W.G.) diameter hypodermic needle along one of the tubes which guides the needle to the sealing ring (on this account the tubes are called guide tubes) and piercing the latter with the needle. The elasticity of the silicone rubber of the ring ensures complete sealing of the hole left by the needle after the latter's withdrawal and, moreover, continues to do so after as

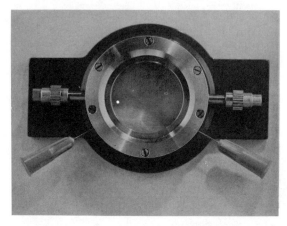

Fig. 5.5. Plan view of Prior chamber mounted on base plate and with gas needles in position. The gas needles must have metal mounts in order to withstand hot-air sterilisation

many repetitions of the process as are likely to be carried out during the life of any one culture. Continuous perfusion of a culture can be set up by using both guide tubes.

The lower part of the housing is provided additionally with two fine radial drillings which lie in the same horizontal plane as that of the guide tubes and, in the original design, are disposed symmetrically about the latter and 100° apart (Fig. 5.4). In the Prior chamber they also lie symmetrically to the guide tubes but between them and 115° apart (Fig. 5.5). A 13 mm ($\frac{1}{2}$ in) long, 0·55 mm (24 S.W.G.) diameter hypodermic needle is passed along each of these drillings and pushed through the sealing ring so that its tip enters the peripheral part of the chamber which acts as the gas space. These short needles, which remain in place throughout the period of cultivation, are equipped with cotton-wool dust filters and serve to prevent pressure changes within the chamber especially when liquid is added to or withdrawn from the central part of the latter via the guide-tube needles. Fitted with suitable connections they afford means for the intermittent or continuous optimum physiological adjustment of the atmosphere in the gas space as well as for the admittance to the latter of experimental gas mixtures. On account of these functions they are called the gas needles.

The chamber, enclosed in a Petri dish, is hot-air sterilised in the assembled condition with the cover-glasses greased and the gas needles, capped with small glass tubes, in place but with the clamping screws adjusted so as not to compress the sealing ring. The screws are fully tightened after sterilisation is completed and the chamber has cooled down. The cotton-wool dust filters, contained in short lengths of vinyl or glass tubing are sterilised separately in the autoclave and attached subsequently to the gas-needle mounts. The chamber housing is then secured to a metal base plate (Fig 5.4) which provides for the mounting and ready movement of the chamber on the microscope stage. The chamber is then ready for the introduction of the cells to be cultured.

The cells, suspended in a suitable medium (see p. 126) at an appropriate concentration (see p. 131) are taken into a small sterile syringe fitted with a 38 mm long, 0·55 mm diameter hypodermic needle, the cap of one of the guide tubes is flamed and removed, the end of the guide tube flamed and the needle passed along the tube, pushed through the sealing ring to enter the chamber and advanced until its tip reaches the centre of the latter. It will then be found that, by virtue of the clearance provided between the guide-tube bore and the needle together with the elasticity of the silicone rubber of the sealing ring, it is possible by manipulating the syringe almost, if not actually, to touch either cover-glass with the tip of the needle (care must be taken to avoid the mischance of scratching the glass with the needle). With the bevel of the needle facing upwards and close to the upper cover-glass a little of the suspension is cautiously discharged and applied to the upper cover-glass to form a hanging drop. The quantity of suspension run in is then gradually increased until the lower surface of the drop makes contact with the lower cover-glass. A little further addition up to a total volume of, usually, about 0·2 ml for the Trevan and Roberts chamber and about 0·25 ml for the Prior chamber may be needed completely to cover the ungreased central areas of both chamber walls and thus to fill the chamber. (It will be noted that the introduction of the suspension occasions no rise of pressure within the chamber because of the communication between the latter and the exterior afforded by the gas needles and their filters). The needle is then withdrawn, the guide-tube end and the cap flamed and the cap securely replaced. The chamber is put into a sterile container and is then ready for incubation to be started.

If a conventional erect microscope is to be used for the observation of the cells the incubation must begin with the chamber inverted so that the cells shall deposit on the upper wall. After 2–3 hours in the inverted position the macrophages will have become adherent and the incubation may be continued subsequently with the chamber the right way up.

When the whole chamber has warmed up to incubator temperature—usually by about the time when the macrophages have become adherent—the chamber is gassed. This process may be intermittent, as in the case of tube cultures in which it is the only method possible, or continuous. If intermittent, the warmed gas mixture is admitted through one of the gas needles and its filter and the existing contents of the gas space—atmospheric air on the first occasion—flushed out through the other gas needle. The gas needle connections are then clamped shut and the incubation continued. If continuous, the gas mixture, before it is led into the chamber, must be fully saturated with water vapour and warmed by passing it through a wash bottle maintained at a temperature a few degrees above that of the chamber. In addition the exit needle should be connected to a tube dipping under water in a shallow vessel in order to provide a visible indication of the gas flow and also to keep constant, in the face of variations of the external atmospheric pressure, the slight difference between the latter and the pressure within the chamber. The gas-needle system when used in this latter way affords a very efficient means of maintaining the optimum physiological atmosphere in the gas space. It may also be used to maintain with equal efficiency experimental alterations of the gas-space atmosphere.

Periodic renewal of the medium and application of various agents to the culture for the purposes of experiment can be carried out easily through a guide-tube needle. Carefully performed these operations need entail no risk of mechanically disturbing the adherent cells. With both guide tubes in simultaneous use and their needles so arranged as just to enter the culture medium, continuous perfusion of the culture with normal medium or media experimentally altered can be set up using slight positive or negative pressure as the motive force. Whichever of the latter methods be employed it is obvious that care must be taken to make certain that no disparity can develop between the rates of inflow and outflow.

Whenever cultivation is terminated the upper cover-glass may be removed from the chamber and the cells adherent to it fixed and stained for microscopy as in the flying-cover-slip technique.

Reference was made earlier to the setting of the guide tubes diametrically opposite to one another in the Prior chamber, whereas they are only 60° apart in the original design, and to the fact that this modification leads to a certain difficulty in use. It is obvious that, when continuous perfusion is being employed, the Prior arrangement is likely to provide a more uniform distribution of the medium to the cells than is the original and it is, therefore, an improvement in this respect. When the chamber housing is mounted on its base plate the guide tubes lie directly over the latter (Fig. 5.5) and doubtless in order to keep the housing as low as possible in the plate, the clearance between the guide-tube caps and the plate is very small. These circumstances render the removal and secure replacement of the caps a little difficult and, more seriously, lead to a very real risk during the flaming of the caps and the guide-tube ends of heating up the base plate and, by conduction through the latter—it is made of an aluminium alloy—as well as directly, the adjacent parts of the chamber housing. This danger can be minimised by the careful use of a small but forceful flame applied in the horizontal plane first from one side and then from the other and directed away from the chamber housing. It will be seen (Fig. 5.4) that in the original design the disposition of the guide tubes and the shape of the base plate eliminate these difficulties.

The foregoing account has been given in some detail in order to draw attention to the many virtues of this chamber—it was, indeed, well-called versatile by Trevan and Roberts in the title of the paper in which they described it. Further information concerning the use of the chamber is given in that paper the instructions contained in which are summarised in a leaflet issued by W. R. Prior & Co. Ltd., which also gives one or two useful additional practical hints.

THE CRUICKSHANK, COOPER AND CONRAN AND STERILIN CHAMBERS

Another very useful and much less complicated chamber is that devised by Cruickshank, Cooper and Conran[30]. The principle of this chamber is basically the same as that of the Trevan and Roberts chamber in that a small volume of culture medium is retained by surface tension in the interval between two parallel transparent walls and surrounded by a gas space.

In the original design (Fig. 5.6) a circular flat-bottomed cavity 25 mm diameter is machined to a depth of not more than 0·75 mm in the central part of one surface of a 5 mm thick piece of Perspex of convenient size (75 × 40 mm)

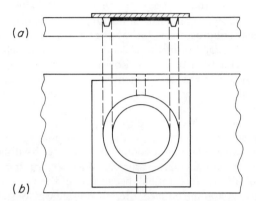

Fig. 5.6. Scale drawings of central part of Cruikshank, Cooper and Conran chamber with cover-glass in place.
(a) Longitudinal section
(b) Plan view with side ducts indicated by dotted lines.

for accommodation on the microscope stage. The peripheral 2·5 mm of the floor of the cavity is then further machined to a total depth of 3 mm to produce an annular recess surrounding the remainder of the cavity floor which thus presents as the flat top of a squat pillar 20 mm diameter. The top of the pillar is polished and, to complete the chamber, the whole cavity is roofed over with a 38 mm square cover-glass which is cemented to the surrounding surface of the Perspex with beeswax*.

The medium is confined by surface tension between the pillar top and the cover-glass and surrounded by the annular recess which provides the gas space. Access to the latter and to the space between cover-glass and pillar top is afforded by two diametrically opposed ducts of 3 mm diameter drilled through the long sides of the Perspex plate.

The Perspex plate is sterilised by boiling, a ring of melted beeswax deposited on the surface of the plate concentrically with the cavity by means of a hollow cylindrical applicator and the cover-glass flamed and laid in place. The wax spreads beneath the cover-glass and should be of sufficient quantity to make a complete seal and to run up to the edge of the cavity.

The cell suspension is introduced via one of the ducts by means of a syringe and hypodermic needle, the chamber gassed and the ducts plugged with short lengths of swab stick which have been soaked in melted beeswax. The chamber

* Paraffin wax is unsuitable for it is permeable to culture media.

is incubated upside down for the first few hours so that the cells may have time to deposit on and adhere to the cover-glass. It may then be turned cover-glass up for microscopy and kept that way during further incubation. Renewal of the medium and the application to the culture of other substances and periodic regassing are done through the side ducts by means of which continuous perfusion can also be carried out. In the latter procedure little rubber bungs through which the needles are passed are fitted into the mouths of the ducts.

The separation from the Perspex of the cover-glass for the fixing and straining of the cells is apt to be a little difficult partly because of the rather hard and brittle nature of solid beeswax, but by the patient insinuation of a safety-razor blade under the corners of the glass the latter can usually be removed unbroken. The difficulty can be lessened by using a mixture of equal parts of histological-grade paraffin wax of low melting point and beeswax instead of the latter alone. This mixture is appreciably more plastic than 'pure' beeswax and appears not to be permeable by the medium.

The advantages of this chamber are its simplicity and the ease of its manufacture in the laboratory workshop so that several may be made and used simultaneously in experiments of a comparative nature. Its disadvantages are threefold. Firstly, compared with the Trevan and Roberts or Prior chamber, the maintenance of sterility is more difficult because, once the chamber has been set up and the cell suspension introduced, there is no safe way of re-sterilising the mouths of the access ducts. The most careful aseptic precautions are necessary, therefore, in the subsequent use of the ducts it is recommended that this operation be carried out under a hood and that the chamber be kept in a sterile Petri dish except when on the microscope. Secondly, Perspex releases certain cytotoxic substances the effects of which can only be countered by the use, which may be undesirable, of high concentrations of serum in the culture medium. Thirdly, the thickness of the chamber floor (4·25 mm*) leads to illumination troubles not wholly to be overcome by the use of a condenser of long working distance.

The first of these drawbacks remains but the latter two have been eliminated in a modification of the original design effected jointly by Dr. Cruickshank, A/S Nunc of Denmark, who manufacture the redesigned version on a commercial scale, and Sterilin Ltd.†, who sell it in this country. The major changes in the modified form are the substitution for the machined Perspex of the original design of an accurate moulding in Nunclon, which is a plastic whose surface is non-toxic and water wettable, and the hollowing out from below of the central pillar so that the chamber floor is reduced in thickness to 1·4 mm (Fig. 5.7) with a resultant improvement in the optical qualities of the chamber. Several minor dimensional changes have also been made. The length and breadth of the plate are reduced to 55 and 35 mm respectively, the latter alteration necessitating the use of cover-glasses 32 mm square. The thickness of the chamber-containing part of the plate remains at 5 mm but a 10 mm length of each end of the upper surface of the plate is lowered by 3·6 mm to provide convenient places for the application of stage clips and, at the same time, avoid the risk of

---

* If thinner Perspex be used the plates tend to warp after several sterilisations.
† Sterilin Ltd., 12 Hill Rise, Richmond, Surrey.

damage by the latter to the chamber top. The chamber dimensions are also slightly changed, namely the overall diameter is reduced to 23 mm and the pillar diameter to 17 mm while the width and depth of the annular recess are increased to 3 mm and 3·6 mm respectively. The chamber depth is standardised at 0·7 mm. The side ducts are reduced in diameter to 1·5 mm and sloped upwards towards the chamber at such an angle that their axes, when produced, meet the rim of the pillar top.

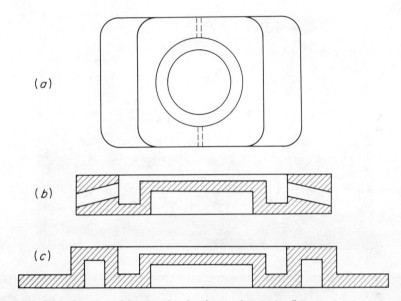

Fig. 5.7. Scale drawings of Sterilin chamber (cover-glass omitted).
(a) Plan view with side ducts indicated by dotted lines. Scale 1:1
(b) Cross section showing hollowed-out central pillar, circumferential gas-space recess and side ducts. Scale 2:1
(c) Longitudinal section showing place for stage clip at each end. Scale 2:1

This modified version is supplied ready sterilised by irradiation and is intended to be used once only and then discarded. Otherwise it is used in the same way as is the original chamber though we have had no experience of using it for continuous perfusion.

THE WASLEY MICROPLATE CHAMBER

A microplate is in effect a collection of small test-tubes or wells moulded in a single piece of plastic (Fig. 5.8). Microplates were originated by Takatsy[31] for the microtitration technique he devised and were used by Takatsy, Furesz and Farkas[32] in their method for the quantitative study of the union of influenza virus and antibody. They were first used as multiple cell-culture chambers by Sever[33] who cultivated in them suspensions of cells for the microtitration of virus and virus and serum reactions in metabolic-inhibition tests. The end points of the titrations were ascertained by means of an indicator (phenol red) incorporated in the medium and there was no microscopical examination of the cells. The titrations were read after incubation for 3 days and to prevent evaporation the contents of each well were covered with a thin layer of mineral oil. Sullivan and Rosenbaum[34], again for virological studies, adapted the method for the culture of cells in monolayers so that cell changes could be observed microscopically.

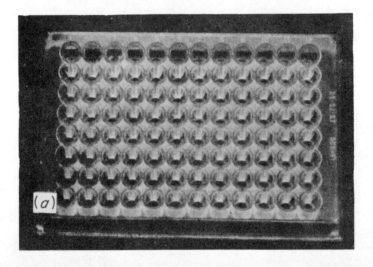

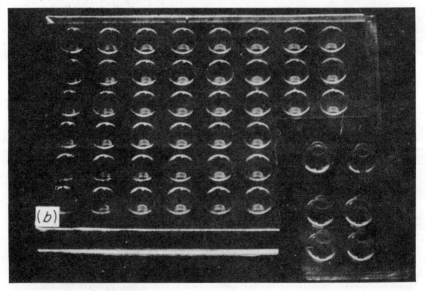

Fig. 5.8. Microplates.
(a) Heavier chamber plate type IF–FB–96. Scale $\frac{2}{4}$:1
(b) Lighter plate for providing plugs type FB–48. Scale $\frac{2}{3}$:1

Whereas the original plates used by Takatsy and those used by Sever were made of Plexiglass and the bottoms of their wells conical and hemispherical respectively, Sullivan and Rosenbaum used plates moulded in a wettable plastic with flat well bottoms on which the cells deposited. These authors also employed a layer of mineral oil to seal the wells during incubation and had of necessity, therefore, to use an inverted microscope for the examination of the cells.

The use of microplates for cell culture in virological work and for many other purposes has become increasingly common and, as a result, a variety of plates with wells of different shapes and sizes has become available commercially. Wasley[35] sought for a means of sealing the wells so that they would be gas-tight and also in such a way that the plates might be inverted and the deposited cells adherent to the bottoms of the wells examined with an erect microscope. He found that the conical-shaped flat-bottomed wells provided in a certain type of plate moulded in a thin plastic could, when cut out of the plate, be utilised to form nicely fitting plugs when pressed into the mouths of the very slightly tapered flat-bottomed wells of another type of plate of much heavier construction. The plates in question are manufactured by the Linbro Co. of Newhaven, Connecticut, U.S.A. and may be purchased in this country from Flow Laboratories Ltd*. They are moulded in a non-toxic plastic and are intended to be used once only and then discarded. The lighter plates from which the plugs are cut are designated FB–48 and the heavier chamber plates IS–FB–96 (Fig. 5.8). The internal diameter of the bottoms of the chamber wells is approximately 5·5 mm and the capacity of each chamber with the plug inserted is just over 0·2 ml.

The wells which are to serve as plugs are cut out with scissors together with sufficient of the surrounding flat part of the plate to leave around each well, after trimming up, an annular rim a couple of mm wide. This rim preserves the rigidity of the plug and facilitates its handling with forceps. In use the plug is placed in the mouth of the chamber well, the forceps laid across its top and the plug pressed home. Since no great pressure is needed to secure a firm and gas-tight joint the plugs are easily removed when necessary. In order to provide adequate clearance between the rims of adjoining plugs, alternate chamber-plate wells only are used in each row. The layout of the wells in the plate is such that all the rows may be used provided that the wells used in any one row are opposite those unused in the adjoining rows. Since the plate contains 8 rows of 12 wells it follows that 48 of the wells can be in simultaneous use.

The chamber plates and plugs are prepared for use by immersion in 70% absolute alcohol in double-glass-distilled water for 30 minutes followed in turn by 4 rinses with double-glass-distilled water, air drying at room temperature in a sterile container and ultraviolet irradiation for 30 minutes. It is as well to prepare an extra supply of plugs so that when, during use, a chamber has to be opened it may be resealed with a fresh sterile plug, to carry out all such operations under a hood and to keep the plates in a sterilised container except when on the microscope.

The quantity of liquid run into each chamber is fairly critical and should be such that none overflows when the plug is inserted and that afterwards there is no more than a very small bubble of air present. Using the microplates specified

---

* Flow Laboratories Ltd., Victoria Park, Heatherhouse Road, Irvine, Ayrshire, Scotland.

a quantity of 0·2 ml. usually fulfils these conditions. If there be excess of air, water vapour condenses during incubation on the under surfaces of the plugs in the form of minute droplets which tend not to coalesce when the plate is inverted for microscopical examination and thus to interfere with the illumination. Further, on inversion of the plate, a large air bubble will rise to the top of the chamber and produce distortion of the microscopical image by refraction effects.

The optical qualities of these microplate chambers are much better than those of culture tubes but, because of their depth (about 6·5 mm) and the thickness (1·5 mm) of their bottom walls, they are naturally inferior to those of the Trevan and Roberts–Prior and the Cruickshank–Sterilin chambers. Nevertheless, correctly filled and used with a condenser of long working distance, they give useful images with objectives of magnifying powers up to 20 diameters.

The complete sealing of the chamber with the inclusion of only a very small volume of air effectively prevents loss of carbon dioxide from the medium, while the quantity of oxygen naturally dissolved in the latter together with that present in the air bubble suffices to meet the requirements of the cells provided that they are not too numerous and that the medium be renewed at intervals not exceeding 3 days.

Macrophages cultured in these chambers remain rounded and unattached to the chamber floors for about 12 hours after their introduction whereas comparable cells cultured in glass-floored chambers, but otherwise under identical conditions, become adherent in 2 to 3 hours. After this initial lag period, however, the cells appear to behave in a perfectly normal way and, renewing the medium at 3 day intervals, have been kept alive for as long as 10 days without developing any microscopically observable abnormality. The delay on the part of the cells in becoming adherent to the chamber floors is possibly associatied with the fact that the particular plastic of which these microplates are made has a non-water-wettable surface.

The Wasley microplate chambers are especially useful when the cells in the source material are sparse, as is sometimes the case when the ones concerned are macrophages, for the small area (7·5 sq mm) of the chamber floor and the relatively large volume of the suspending liquid result in a concentration of the cells as they deposit. They also provide a simple means whereby a number of identical cultures may be set up and the morphological changes effected by various reagents in varying concentrations conveniently studied.

MEDIA SUITABLE FOR THE CULTURE OF MACROPHAGES

The subject of media in general is dealt with fully in Chapter 1. However, a brief repetition of the basic principles may not come amiss for the medium is the most important single factor concerned in the artificial culture of cells and tissues.

The qualities and functions of the medium are several: it must provide initially and maintain subsequently the hydrogen-ion concentration and osmotic pressure requisite to prevent the immediate and later destruction of the cells; it must provide the cells with all the substances, such as various mineral salts, amino acids, carbohydrates and vitamins, which they cannot synthesise for themselves and are essential for their continued survival; it must be pervious to oxygen and carbon dioxide.

DEFINED AND NATURAL MEDIA

It is obvious that the study of the biochemical and other aspects of cell life would be at least somewhat simplified were it possible to cultivate cells in a wholly artificial medium made up of a mixture of pure chemical substances of known composition. Such 'defined' media have been compounded and used successfully for the artificial cultivation of plant cells. Up to the present time, however, with the exception of one or two special strains of cells which have been cultivated through many generations, no such accurately defined medium has been devised that will support the majority of animal cells. The latter require either a wholly natural medium, such as plasma, or an artificial medium supplemented by the addition of some natural product such as serum. The use of supplemented defined media is obligatory for the cultivation of cells in monolayers since these media are liquid whereas plasma, unless treated with undesirably high concentrations of an anticoagulant, promptly clots on making contact with the water-wettable surfaces of the culture vessel and so prevents the necessary initial deposition of the cells.

THE SUPPLY OF OXYGEN AND CARBON DIOXIDE

Most normal adult cells, including macrophages, require a supply of oxygen for their continued survival and even those which are able wholly to satisfy their energy needs without oxygen by glycolysis need certain of the by-products of the Krebs cycle for the synthesis of various cell components. Most cells, however, in culture as in the intact animal, make use of both respiration and, though oxygen be present, glycolysis for the metabolism of carbohydrate.

In tube cultures the oxygen is available to the deposited cells by diffusion through the overlying medium from the large volume of air which occupies the rest of the tube. In culture chambers, such as the Trevan and Roberts–Prior and Cruickshank–Sterilin chambers, the gas space surrounding the 'edge' of the medium forms a relatively large volume compared to that of the medium and, even when completely sealed, such chambers contain sufficient oxygen to meet the needs of the cells during the periods between medium changes. In the case of the Wasley sealed microplate chamber the cells are usually present in small numbers and so, as was mentioned in the description of that chamber, are able to gain sufficient oxygen from that present in the small air bubble and that naturally dissolved in the medium to meet their requirements between medium changes.

For their optimum growth and activity different kinds of cells require correspondingly different tensions of oxygen in the culture medium and, therefore, in the atmosphere to which the latter is exposed. Happily, the optimum oxygen tension in the gas in contact with media in which the macrophages of post-embryonic mammals are being cultured appears to be that of the normal atmospheric air.

The presence of carbon dioxide in the medium is essential for some cells and desirable for all. Produced by the respiration of the cells it dissolves in the medium to give rise to bicarbonate ions. The latter exert their powerful buffering action not only in the medium but also, possibly, actually within the cells. There is some evidence, too, that carbon dioxide plays a part within the cells in the Krebs cycle.

When cells are cultured in a vessel containing a gas space some of the carbon dioxide produced is lost to the latter. Clearly the larger the gas space in relation to the volume of medium and the number of cells present the greater will be the loss which may be minimised, therefore, by making the gas space as small as possible and sealing the chamber—a state of affairs automatically achieved by the Wasley microplate chamber. Otherwise some means must be employed to maintain a carbon dioxide tension of roughly 5% in the atmosphere to which the medium is exposed. This may be carried out very simply by, from time to time, gently blowing expired air through a suitable filter into the gas space or, much more elaborately, as can be done with chambers of the Trevan and Roberts-Prior type, by passing through the gas space a continuous flow of gas of the required composition.

THE USE OF ANTIBIOTIC SUBSTANCES

There is no doubt that the advent of antibiotics and the discovery that some of them may be used in culture media in bactericidal concentrations without apparent harmful effect on the cultured cells or tissues have been largely responsible for the great increase in the practice of artificial cell and tissue cultivation which has developed during the last couple of decades. It must be remembered, however, that all the early work, and there was a great volume of it, was done without the aid of these substances and was successful as the result of the careful sterilisation of materials and the painstaking application or ordinary aseptic methods. Though, nowadays, antibiotics enter into the composition of most culture media this is no reason for relaxing the standard aseptic measures and it is very necessary not to hide behind the protection afforded by these substances against microbic infection, for to do so is to run the risk of developing generally sloppy working methods—antibiotics should not be added to media to kill organisms present as the result of careless work but only to act as a defence against subsequent accidental contamination. Furthermore, it must be borne in mind that antibiotics, as indicated by their very name, may not be entirely without effect on the cells cultured in media containing them.

THE COMPONENTS OF A SUITABLE MEDIUM

The medium now to be described has been found to be most satisfactory for the cultivation of macrophages taken from those animals listed on page 109. Its components are as follows:

(a) The defined part

This may be either Eagle's Medium or Medium No. 199 of Morgan, Morton and Parker. Both these defined media are based on a balanced salt solution and contain phenol red to act as an indicator of pH. They may be purchased already sterilised by membrane filtration from Wellcome Reagents Ltd. * Both Eagle's Medium (the basic version is required) and Medium 199 should be bought in 10× concentrated form without sodium bicarbonate and antibiotics. Their respective code numbers† are TC 38 and TC 22. In this concentrated form both will keep in satisfactory condition for periods up to 3 months if stored in the refrigerator at about 4°C (permissible range 2–10°C). The concentrated solutions are diluted 1 in 10 by volume with sterile double-glass-distilled water‡.

(b) The natural-product part

Calf serum is very satisfactory for mouse macrophages but homologous serum is required for human and autologous serum (see p. 135) for rabbit cells. The calf serum may be bought from Wellcome Reagents Ltd. and the No. 2 unheated variety, code numbers TC 42, TC 43 and TC 44 according to the quantity purchased, should be used. Stored under the same conditions as those used for the defined media it will keep for periods up to one year. Human serum can usually be provided by the haematology department of any large hospital.

(c) Heparin

This substance is used and in a very low concentration to prevent the cells from sticking together during the period before and while their deposition is taking place. It is conveniently obtained from Weddel Pharmaceuticals Ltd § as Heparin (Mucous) Injection B.P. (without preservative) in 5 ml glass ampoules at a concentration of 1000 IU/ml. It keeps well in an ordinary refrigerator. If an ampoule be opened and its contents only partly used it should be kept thereafter in a sterile container.

(d) Antibiotics

Penicillin and Streptomycin are used in combination and will deal respectively with Gram + ve and Gram − ve bacteria. 1 000 000 units of penicillin and 1 000 000 µg (1·0 g) of streptomycin are dissolved in 100 ml of water to make a convenient quantity of stock solution which is stored in the ordinary refrigerator.

---

* Wellcome Reagents Ltd., Wellcome Research Laboratories, Beckenham, Kent.
† Wellcome Price List of Laboratory Diagnostic Reagents and Materials, July, 1969.
‡ Only water so prepared should be used for making up the various solutions required.
§ Weddel Pharmaceuticals Ltd., 24 West Smithfield, London, E.C.1.

(e) Bicarbonate buffer solution

4·4 g sodium bicarbonate and 1·0 mg phenol red are dissolved in 100 ml water and filtered carbon dioxide bubbled through the solution until it becomes a very faintly pink-tinged amber colour. This solution is put into bijou bottles in quantities of 5 ml per bottle and the bottles securely capped and sterilised in the autoclave.

PREPARATION OF THE MEDIUM FOR USE

The reagents are put together under aseptic conditions in the following proportions and order:

(a) Diluted defined medium   90 ml

(b) Serum (calf or as required)   10 ml

(c) Heparin solution   1 ml (this gives a concentration of approximately 10 IU/ml of the medium)

(d) Antibiotic solution   1 ml (this gives a concentration of approximately 100 units of penicillin and 100 μg of streptomycin/ml of the medium)

(e) Bicarbonate buffer solution   This is added drop by drop until the yellow colour of the mixture darkens to a very faintly pink-tinged amber when the pH of the medium will be about 7·2.

PRACTICAL PROCEDURES FOR OBTAINING AND CULTURING MACROPHAGES

In the following pages methods for collecting macrophages from the mouse and human peritoneal cavities and the rabbit lung and of setting up cultures of the cells are described in detail.

THE MOUSE PERITONEUM

As was pointed out in the discussion on sources of supply this is probably the most frequently used source at the present time. Mice of any or no particular strain and of either sex are all equally suitable. They should, however, be about 6 weeks old for the yield of cells from mice of this age is greater than that from either younger or older animals and is large enough for most purposes without the necessity for the previous intra-peritoneal injection of foreign substances.

The principle of the method is very simple: under aseptic conditions a suitable quantity of medium together with a smaller volume of sterile air is injected into the peritoneal cavity of the recently killed animal; the abdominal wall is then subjected to gentle massage which distributes the medium throughout the cavity and simultaneously causes the release of macrophages from the serous surfaces of the latter thus producing a suspension of the cells in the medium; as much of the suspension as possible is then withdrawn. Usually the volume of cell suspension recoverable is about 85% of the injected medium.

In practice 3·5 ml of medium is a suitable quantity for use with mice of the age specified so that each mouse may be expected to yield about 3 ml of cell suspension. The volume of suspension required for a particular experiment obviously governs the number of mice needed and is itself dependent upon the number of cultures to be set up and on the type of culture vessel to be used. 1 ml is needed for a simple or Leighton tube, about 0·2 ml for the Trevan and Roberts chamber, about 0·25 ml for the Prior and Cruickshank, Cooper and Conran Chambers, 0·16 ml, for the Cruickshank–Sterilin chamber and 0·2 ml for each Wasley sealed microplate chamber.

The suspension is transferred immediately after withdrawal to a sterile flask in which, when several mice have to be used, the individual contributions are pooled. It is very important to keep the flask chilled in order to prevent the macrophages from sticking to the glass and advantageous also to chill the medium before injecting it. These purposes are simply served by standing the flask and medium container in a bowl in which ice cubes are melting. It is also important to work quickly for undue delay in collecting the cells after the animal has been killed leads to a reduction in the yield. If more suspension is needed than can be obtained from one mouse it is inadvisable, therefore, to kill all the animals at the same time but better to kill them as required or at most in batches of not more than 3 or 4.

The presence of the air injected with the medium—about 1·5 ml is the optimal quantity—enhances the effect of the massage on the distribution of the medium and, apparently, on the release of the macrophages. By increasing the volume of the peritoneal cavity it also facilitates displacement of the intestines to one side so that the suspension can be more readily and completely withdrawn.

METHOD

1. Preparation of the animal

(a) The mouse is killed by dislocating the neck, pinned out by the limbs in the supine position and the fur of the ventral and lateral surfaces of the trunk and neck well swabbed with 70% alcohol. These steps are preferably carried out by an assistant. Meanwhile the operator should wash his hands thoroughly in running water using some mildly antiseptic soap such as Cidal, which contains Hexachlorophane, and, after thoroughly rinsing, dry them with a sterile towel.

(b) With the hind end of the animal towards the operator the skin in the midline of the abdomen and just foreward of the symphysis pubis (the penis in a male) is grasped with toothed forceps and pulled upwards so as to form a little midline fold which is cut across with scissors. The forward edge of the resulting short transverse skin incision is then grasped with the forceps and pulled upwards from the underlying deep fascia, one blade of the scissors inserted beneath the skin and a midline incision made in the skin and extended over the abdomen, thorax and neck almost to the mouth. The lower blade of the scissors is kept slightly raised as the incision is being made so as to avoid injury to the underlying tissues. Mayo-type scissors with their blunt-ended narrow blades are very suitable for this purpose. Each edge of the longitudinal incision is then gripped a little rearward of its midpoint between the thumb and forefinger of one of the hands and both edges simultaneously pulled laterally and, at first, slightly upwards so that the skin is stripped from the underlying deep fascia of the trunk to expose both flanks and both groins. The flaps of skin so produced are then pinned down under slight tension. By these means all but the midline region of the deep fascia covering the muscles of the ventral and lateral aspects of the abdominal wall is exposed aseptically.

2. Injection of the medium and air

The injection is made through the abdominal muscles at a point a little forward of the left groin and lateral to the potentially contaminated region of the midline. A 5 ml syringe fitted with a 0·45 mm diameter (26 S.W.G.) hypo-dermic needle 16 mm ($\frac{5}{8}$ in) long is used. 1·5 ml of sterile air are drawn into the syringe from the container in which the latter was sterilised followed by 3·5 ml of the chilled medium.

With the syringe held so that the long axis of the needle is at once approximately parallel to the midline of the animal and to the surface of the board to which the latter is pinned, the tip of the needle is just entered into the external muscle layer of the abdominal wall. This latter step is most easily carried out when the bevel of the needle tip faces away from the surface of the muscle. With the tip of the needle thus engaged and without changing their direction, the syringe and needle are rotated through 180° and the needle pushed slowly through the remaining layers of the muscle while simultaneously applying slight pressure to the plunger of the syringe. Because of the natural bulging of the abdominal wall the needle passes obliquely through the latter with the plane of its bevel practically parallel to that of the parietal peritoneum. On perforation of the parietal peritoneum by the needle—the moment of this occurrence is readily recognisable for the thinness and degree of translucence of the musculature are such that the displacement of neighbouring loops of gut consequent upon the resulting inflow of medium is quite clearly visible—the whole of the contents of the syringe are slowly discharged meanwhile the needle is pushed in a further millimetre or so. By these means the risk of a part of the gut or other viscus being picked up by the needle tip and receiving the injection is wholly eliminated. The injection proceeds to the accompaniment of gradually developing distension of the abdomen which on completion of the process is rendered markedly protuberant. The needle is then withdrawn whereupon those parts of the muscle layers of the abdominal wall previously penetrated by the needle, and, while the latter was in place, held by it in the positions relative to one another which obtained before the abdomen was distended, slide into new positions consonant with the stretched state of the abdominal wall and so disrupt and tend to seal the needle track. The greatly raised tension in the muscles and the length of the needle track consequent upon its obliquity combine to enhance this sealing action and usually make it complete.

3. Massage of the abdomen

This process consists in lightly but quite rapidly tapping the abdominal wall for not less than one minute in a region to the left (the animal's left) of the most protuberant part. The tapping is done with the tip of the right fore or middle finger with the rest of the operator's hand held well to the animal's left.

4. Withdrawal of the cell suspension

A fresh 5 ml syringe fitted with 0·8 mm diameter (21 S.W.G.) hypodermic needle 38 mm ($1\frac{1}{2}$ in) long is used. The needle is passed into the peritoneal cavity at a point just rearward of the most protuberant part of the right flank. The syringe and needle are positioned, the needle tip inserted into the muscle and the syringe and needle rotated exactly as was done for the injection of the medium and air. Before the needle enters the peritoneal cavity, however, the syringe and needle are drawn sideways to pull out the flank and so to form a peritoneal pocket into which the suspension drains and accumulates to the exclusion of most of the loops of gut (Fig. 5.9). The needle is then passed into the pocket and the suspension withdrawn—usually without interruption caused by occlusion of the needle bevel by gut—and transferred to the chilled flask.

5. Estimation of cell concentration

When the requisite amount of suspension has been collected it is as well to take a sample for the purpose of making a cell count in case the suspension be too concentrated. This is done by means of an ordinary blood-counting chamber. A drop of the shaken-up suspension, undiluted and unstained and of a size suited to the particular chamber used is placed in the latter, the cells rendered visible by closing down the iris diaphragm of the microscope and all of them, irrespective of type, counted. A concentration of about 1 000 000 cells/ml is usual and of these cells about one third will be macrophages the remainder comprising macrophage precursors, lymphocytes, eosinophils and mast cells. If the concentration prove to be appreciably higher the suspension should be diluted appropriately with chilled medium. The suspension is then ready for use.

6. Setting up the cultures

The suspension is run into the culture vessels in quantities suited to the particular type of vessel used (see page 129) and in accordance with the directions given in the descriptions of the various types. The vessels are then placed in the incubator at 37°C.

After a couple of hours most of the cells present in the suspension will have deposited on the culture-vessel floors and, if the latter be of glass or a water-wettable plastic, the macrophages, but none of the other cells present, will have adhered to them. In the case, however, of chambers whose floors are not water wettable, for example the original Cruickshank, Cooper and Conran and the Wasley microplate chambers, the adhesion of the macrophages does not occur until about 10 hours after their deposition.

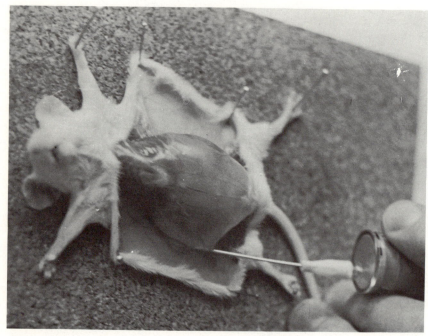

Fig. 5.9. Abdomen of mouse distended with medium and air and with needle and syringe in place and flank drawn out in readiness for withdrawal of suspension

Having allowed the requisite time for the macrophages to become adherent, the original medium is washed out with fresh medium which has been previously warmed to 37°C by which simple means the cultures are freed of most of the contaminating cells. The culture vessels may now be inverted for microscopical examination of the macrophages and experimental work on the cultures started.

HUMAN PERITONEAL DIALYSATE

When available locally this is an excellent source for not only are the macrophages obtained from it human but the yield of them, though variable, is usually good and, as previously mentioned, the number of cells of other types* present in it is generally small. These advantages are, however, offset to some extent by the possibility, already referred to on page 108, that the dialysate may be infected with the causative agent of homologous-serum jaundice. Though this agent does not, so far as we are aware, have any deleterious effect on the macrophages, the possibility of its presence does necessitate the taking by all persons handling the dialysate of certain precautions against accidental infection which must be rigorously carried out. It is certain that this agent may enter the body through minute cuts in the skin and so rubber or other protective gloves must always be worn when handling the dialysate and all vessels and other articles that have been in contact with it. There is some evidence that infection may also be acquired via the alimentary tract and this possibility makes it imperative that a well-fitting mask be worn especially when containers and other equipment that could be infected are being washed up, for in this process, apart from ordinary splashing, innumerable fine droplets of the washing-up and rinsing waters come to be suspended in the ambient air and may well be taken into the unprotected nose and mouth. An additional precaution against this latter risk is afforded by soaking the articles in a 5% solution of Chloros (I.C.I.) in water for $\frac{1}{2}$–1 hour before washing and very thoroughly rinsing them.

It will almost certainly—and very properly—be insisted upon that the collection of the dialysate from the patient be made by one of the clinical staff and the person concerned will need, therefore, to be provided with suitable receptacles and instructed in their use. Flat-bottomed bottles of 500 ml capacity equipped with screw caps fitted with silicone-rubber liners so that they may be sterilised in the hot-air oven are very suitable and a supply of these should be provided and arrangements sought for them to be stored in a refrigerator at 4°C or thereabouts at the place where the treatment is carried out. The dialysate should be run directly from the patient into the chilled bottles (with, of course, full aseptic precautions) so that it may be cooled quickly in order to eliminate loss of macrophages by reducing the ability of the latter to adhere to the bottle walls. A volume ranging between 0·5 and 2 litres may be expected from each dialysis the amount in a particular case obviously being dependent on the volume of the preceding intraperitoneal injection which has to be varied according to the condition of the patient. At each treatment session the patient is usually subjected to several dialyses and of these the dialysate from the second generally gives the most satisfactory yield of cells. The bottles containing the dialysate are taken immediately to the laboratory in an insulated box or, if some distance has to be travelled, in a portable ice-box so that the cooling process may continue. It is a great help if the clinical staff be kind enough to give forewarning of the time when a supply of dialysate is likely to be available so that all may be made ready at the laboratory for its reception.

On arrival at the laboratory the dialysate is examined by the naked eye. It should be clear and palely straw coloured though a specimen that is but slightly turbid is always worth proceeding with but one which is blood-stained should be discarded. The details of the subsequent procedure are as follows (the volumes given have been found to be suitable for dealing with 1 litre of a

---

* Erythrocytes, mast cells and desquamated serosal cells.

dialysate from which the yield of macrophages is that obtained from the average specimen, or thereabouts—the volumes are varied proportionally for use with larger or smaller quantities of dialysate):

1. The dialysate is centrifuged at 800 rev/min for 15 minutes at a temperature not above $+10°C$. For this purpose a refrigerated centrifuge is desirable but an ordinary centrifuge is satisfactory provided the buckets and tubes be well cooled before use.
2. The supernatant liquid is poured off and the deposit washed into fresh chilled centrifuge tubes with 100 ml of Hanks' balanced salt solution. The latter may be bought from Wellcome Reagents Ltd. in $10\times$ concentrated form without bicarbonate (code no. TC56 in Wellcome Price List of Laboratory Diagnostic Reagents and Materials, July, 1969). After dilution the hydrogen-ion concentration of the solution is adjusted by the addition of bicarbonate buffer solution as in preparing medium (see page 128).
3. The resulting suspension is centrifuged as in (1) above, the supernatant liquid poured off and the deposit suspended in 10 ml of culture medium. The latter is made up with human serum which for most experimental purposes may come from blood of any group.
4. A sample of the suspension is taken for making a cell count which should be between 300 000 and 500 000 nucleated cells/ml. If necessary the concentration in the rest of the suspension is adjusted by suitable dilution if too high or if too low by recentrifuging and resuspending the cells in an appropriately smaller amount of medium.
5. The suspension is then run into culture vessels (see page 129 for the quantities to be used in the various types of vessel) and incubation at $37°C$ started.
6. After sufficient time has elapsed for the cells to deposit on and the macrophages to adhere to the vessel wall (2–3 hours when the surface is of glass or a water-wettable plastic and about 10 hours when the surface is not water-wettable) the original medium is washed out with fresh warmed medium. By this simple means most of the cells other than macrophages are disposed of.

THE RABBIT LUNG

We have had no experience of using this source but it would appear that it has three very advantageous qualities: firstly, the method of obtaining the macrophages (alveolar phagocytes), which was devised by Myrvik, Leake and Fariss[28], is basically simple and consists in washing the cells out of the lungs via the trachea with a balanced salt solution; secondly, the yield of macrophages is high—it averages about 12 000 000 cells from the lungs of one animal; thirdly, the only living rabbit cells of other types washed out with the macrophages are polymorphonuclear leucocytes, and these are outnumbered by the macrophages by more than 100 to 1 and die off in a day or so in culture, and very rarely lymphocytes most of which also die quickly. These very good characteristics are somewhat offset by three disadvantages of which two are minor but the third is such as might render macrophages derived from this source unsuitable for some experimental purposes: firstly, even in health bacteria are practically always present in the trachae and larger bronchi so that the use of antibiotics in the culture medium is obligatory and an essential part of the technique not merely a safeguard against accidental infection; secondly, because the sera of some rabbits are toxic to the cells of other animals including other rabbits it is safer to use autologous rather than homologous serum for the natural component of the culture medium and this adds a little complication to the technique; lastly, alveolar phagocytes, at any rate those of the rabbit, contain considerable quantities of lysozyme whereas there is little or none of this substance in their counterparts derived from the peritoneum[36] and, probably, other extra-pulmonary sites—obviously this peculiarity would at least have to be taken into account in any experimental study of the action of these alveolar macrophages on bacteria.

The procedure of Myrvik, Leake and Fariss[28] is as follows:

1. Healthy rabbits weighing from 1·8 to 2·2 kg are used—the authors used New Zealand white rabbits but doubtless other varieties would give equally good results. Each animal is killed by the rapid injection of about 40 ml of air into a marginal ear vein.

2. The chest is opened, the surrounding structures dissected away from the anterior end of the exposed part of the trachea and the latter clamped shut with an artery forceps, to prevent ingress of blood, and divided in front of the clamp. The still-clamped posterior part of the trachea, the lungs and the heart are then removed *en masse*. The heart is cut away taking care to avoid injuring the lungs and bronchi in the process.

3. The preparation is swabbed free of blood with warmed physiological saline, gently blotted with gauze mops to remove excess of the saline and then suspended from some convenient support by means of a second artery forceps so applied to the outside of the trachea as to secure a firm grip without either perforating the wall or occluding the tracheal lumen.

4. The trachea is unclamped and Hanks' balanced salt solution run in until the lungs are expanded, 30–40 ml of the solution usually suffice. It is advantageous to run the solution directly into each main bronchus in turn for this ensures uniform distension of all the lobes. The trachea is reclamped, the lungs gently massaged and, after removing the clamp, drained into a suitable vessel. This process is repeated once more.

5. The combined washings are centrifuged at 1500 rev/min for 20 minutes, any hazy layer that may be present between the clear supernatant liquid and the deposited cells—consisting of mucus and cell debris—is carefully removed with a capillary pipette, the rest of the supernatant liquid poured or pipetted off and the deposit suspended in culture medium containing 25% of autologous serum. The suspension is then transferred to the culture vessel and incubation at 37°C begun.

Though, as already stated, we have not used this source of macrophages and have, therefore, no practical experience of the above method, a few points arise in connection with the latter, comment upon which in the light of other experience of this type of work may not, perhaps, be out of place.

In the interest of avoiding the risk of unnecessary microbial contamination of the final product it would be advantageous to carry out the whole procedure under aseptic conditions including the use of sterile air for killing the animal.

It would seem wise to warm the balanced salt solution to 37°C before running it into the lungs in order to avoid the possible constricting effect of a cold douche on the bronchial and especially the bronchiolar musculature. The receptacle into which the washings are discharged should, however, be chilled to prevent adhesion of the macrophages to its walls. The centrifuging of the washings should be done under cold conditions either in a refrigerated centrifuge or in an ordinary machine with the buckets and tubes chilled before use.

The rather high speed and somewhat lengthy duration of centrifuging quoted in (5) above are part of a standardised technique that was used for estimating the yield of cells by measurement of the packed-cell volume. There seems to be no reason, however, for not assessing the yield by counting the cells present in a sample of suspension in the way recommended for peritoneal macrophages nor, perhaps, reducing the speed of the centrifuging with consequent diminution of the liability of the cells to mechanical injury.

If it be assumed that the average yield from one animal is 12 000 000 cells of which 99% are macrophages then the volume of medium required to suspend these cells at a concentration of 300 000 to 500 000 macrophages/ml would be of the order of 30 ml which would allow of the setting up of a large number of cultures.

In regard to the supply of autologous serum it would seem reasonable to draw off say 50 ml of blood from the marginal ear vein before injecting the air to kill the animal. This volume of blood will yield about 30 ml of serum which should be ample for making up all the medium needed for the initial setting up and subsequent refeeding of the cultures.

After setting up the cultures and allowing time for the macrophages to adhere to the culture-vessel walls the original medium should be washed out with warmed fresh medium thereby ridding the culture of any non-adherent matter such as desquamated epithelial and other cells.

It may be of interest to mention that Heise, Han and Weisser[37] have successfully applied this method to guinea-pig lungs.

REFERENCES

1. METSCHNIKOFF, E. (1833 and 1884) cited by CAMERON, R., in *Pathology of the Cell*, Oliver and Boyd, Edinburgh, 196 (1952)
2. ASCHOFF, L. and KIYONO, K. (1913) cited by CAMERON, R., in *Pathology of the Cell*, Oliver and Boyd, Edinburgh, 205 (1952)
3. RANVIER, L. (1899) cited by JACOBY, F., in *Cells and Tissues in Culture*, Academic Press, London, **2**, 27 (1965)
4. DEL RÍO HORTEGA, P., in PENFIELD's *Cytology and Cellular Pathology of the Nervous System*, Hoeber, New York, **2**, section 10, 481 (1932)
5. KUPFFER, C. VON (1898) cited by CAMERON, P., in *Pathology of the Cell*, Oliver and Boyd, Edinburgh, 205 (1952)
6. ASCHOFF, L., *Lectures on Pathology*, Hoeber, New York, chapter 1 (1924)

7. CAPPELL, D. F., *J. Path. Bact.*, **32**, 595 (1929)
8. CAPPELL, D. F., *J. Path. Bact.*, **33**, 429 (1930)
9. DOWNEY, H. (Ed.) *Handbook of Hematology*, Hoeber, New York, **I.** (1938)
10. JACOBY, F., in *Cells and Tissues in Culture*, Ed. WILLMER, E. N., Academic Press, London, **2**, chapter 1 (1965)
11. NELSON, D. S., *Macrophages and Immunity*, Vol. 11 of *Frontiers of Biology*, North Holland Publishing Co., Amsterdam (1969)
12. DEL RÍO HORTEGA, P. and DE ASÚA, F. J. (1921) cited by MARSHALL, A. H. E., *J. Path. Bact.*, **58**, 729 (1946)
13. WEIL, A. and DAVENPORT, H. A., *Archs Neurol. Psychiat., Lond.*, **30**, 175 (1933)
14. MARSHALL, A. H. E., *J. Path. Bact.*, **58**, 729 (1946)
15. CARREL, A., *C.r. Séanc. Soc. Biol.*, **94**, 1345 (1926)
16. TONER, P. G. and CARR, K. E., *Cell Structure*, Livingstone, Edinburgh, 48 (1968)
17. CARREL, A. and EBELING, A. H., *J. exp. Med.*, **36**, 365 (1922)
18. MALLUCCI, L., *Virology*, **25**, 30 (1965)
19. FRANZL, R. E., Personal Communication to GESNER, B. M. and HOWARD, J. G., in *Handbook of Experimental Immunology* edited by WEIR, D. M., Blackwell, Oxford, 1022 (1967)
20. ARGYRIS, B. F., *J. Immun.*, **99**, 744 (1967)
21. SUTER, E., *J. exp. Med.*, **96**, 137 (1952)
22. HIRSCH, J. G., *J. exp. Med.*, **103**, 589 (1956)
23. FISHMAN, M., *J. exp. Med.*, **114**, 837 (1961)
24. COHN, Z. A. and WIENER, E., *J. exp. Med.*, **118**, 991 (1963)
25. LUCKÉ, B., STRUMIA, M., MUDD, S., McCUTCHEON, M. and MUDD, E. B. H., *J. Immun.*, **24**, 455 (1933)
26. VOLKMAN, A., *J. exp. Med.*, **124**, 241 (1966)
27. BISHOP, D. C., PISCIOTTA, A. V. and ABRAMOFF, P., *J. Immun.*, **99**, 751 (1967)
28. MYRVIK, Q. N., LEAKE, E. S. and FARISS, B., *J. Immun.*, **86**, 128 (1961)
29. ROBERTS, D. C. and TREVAN, D. J., *Jl R. microsc. Soc.*, **79**, 361 (1961)
30. CRUICKSHANK, C. N. D., COOPER, J. R. and CONRAN, M. B., *Expl Cell Res.*, **16**, 695 (1959)
31. TAKATSY, G. (1950) cited by SEVER, J. L., *J. Immun.*, **88**, 320 (1962)
32. TAKATSY, G., FURESZ, J. and FARKAS, E. (1954) cited by SULLIVAN, E. J. and ROSENBAUM, M. J., *Am. J. Epidem.*, **85**, 424 (1967)
33. SEVER, J. L., *J. Immun.*, **88**, 320 (1962)
34. SULLIVAN, E. J. and ROSENBAUM, M. J., *Am. J. Epidem.*, **85**, 424 (1967)
35. WASLEY, G. D., *J. med. Lab. Technol.*, **26**, 134 (1969)
36. MYRVIK, Q. N., LEAKE, E. S. and FARISS, B., *J. Immun.*, **86**, 133 (1961)
37. HEISE, E. R., HAN, S. and WEISER, R. S., *J. Immun.*, **101**, 1004 (1968)

SIX

Sub-zero cell storage

JOHN MAY
The Lord Rank Research Centre, High Wycombe, Bucks.

HISTORICAL AND GENERAL SURVEY

In 1776 Spallanzani[1] made one of the earliest published records of the use of 'cold' to facilitate the observations made of cells. On this occasion he conducted microscopy observations of stallion spermatozoa which had previously been chilled with snow. He records his observations as follows:

'The same effect was produced by Snow as by Winters cold, that is in fourteen minutes it made spermatozoa motionless; although when exposed to the heat of the atmosphere they continued to move for seven hours and a half. But an accident that happened in this experiment afforded new intelligence and divested me of prejudice. Observing that the vermeculi had become motionless I took the glass from the Snow and left it exposed to the air. An hour after, by chance observing the Semen with the microscope I was astonished to find all of the vermeculi re-animated, and in such a manner as if they had just come from the Seminal vessels. I then saw that the cold had not killed them, but had reduced them to a state of complete inaction. I replaced them in the Snow, and in three quarters of an hour took them away. These are the phenomena that I observed.

In a few minutes their vivacity relaxed, and the diminution increased until they lost progressive motion, and retained only that oscillation which likewise ended in a few minutes more. Exactly the reverse was observed when they passed from the cold of the Snow to the heat of the atmosphere. The first motion was that of oscillation; the body and then the tail began to vibrate longitudinally from right to left; then the motion was communicated to the whole vermicule, and in a short time progressive motion had begun'.

No preservatives were used in his experiments, the objects of which were to reduce cell motility and thereby enable observations to be made more easily with the little developed microscopes available to him at that time.

Some one hundred years later more of these early workers, Prévost 1840, de Quatrefages 1853, Mantegazza 1866[2], and Scheuk 1870, experimented with sperm exposed to reduced temperatures and recorded their findings.

Mantegazza was interested also in the preservation of human spermatozoa. This work followed his observations that these cells showed some resistance to freezing and thawing. At this time also he showed some interest in the application of these early techniques to farm animals, for what is now known as artificial insemination.

In 1900 it was shown that many biological 'organisms', i.e. seeds, bacteria and some other complex biochemical substances were able to survive storage at sub-zero temperatures in liquid air. The later effect of these observations was a move towards the preservation of food under similar conditions. The results of the particular line of research and development is very familiar to us all now.

In 1940 Shettles[3] showed that human spermatozoa were resistant to sudden variations in temperature. He treated undiluted human seminal fluid, sealed in micro-tubes, at low temperatures of between −79°C and −196°C, finding that some 10% of these treated cells survived.

Research in the field of sub-zero storage has increased as the areas of observation have extended since the 1940s. In 1945 Parkes[4] showed that the survival rate of stored cells could be increased when larger volumes of the specimens were frozen. However, the study of spermatozoa in far greater detail was made possible following the discovery that glycerol offered some protection to cells in sub-zero storage under certain conditions.

Reference should be made to Sherman[5] for a complete review of the position at that time. One of the major advances of the decade was that of Polge, Smith and Parkes[6] in 1949. This work, again carried out on spermatozoa, demonstrated that the addition of glycerol to cell suspensions greatly helped their survival following storage at −79°C.

The classical and valuable work of Luyet[7] in 1951 was another most important contribution to the knowledge of the subject. This was followed by the work of Lovelock[8] during the period 1953 to 1959. He also demonstrated that the rise in the concentration of electrolytes was a major cause of damage to cells during freezing and that this damage reached high proportions when the concentration took place within or near to the cells.

One other contribution to the theory of the freezing of cells is that made by Merryman[9], when he outlined in great detail the mechanics of the freezing processes, and later in 1957[10] work on the wider foundations of the theories was continued and published by Rey[11, 12].

Sub-zero cell storage has been found to have very wide applications in the field of virology and many publications stem from this area of research and development. Some of the major advances in virology have often been made as a result of the combination of the techniques of cell and tissue culture and sub-zero storage. A further major contribution was by the publication of a text on the subject by Smith[13] in 1961.

The period 1949 to 1960 can be called the 'glycerol period', for this was the preservative that was in general use at that time. Towards the end of this period however, a substance with a quite different chemical structure came into the picture, dimethyl sulphoxide (D.M.S.O.).

There is not a general acceptance of the claims that D.M.S.O. is preferable to glycerol as a preservative at this time. The results of long term storage trials using both chemicals will possibly show that each has its place as the preservative for specific cells. The theory supporting the use of dimethyl sulphoxide is outlined by Farrant[14]. Further work on the use of preservatives, other than glycerol, was carried out by Nash[15] and published in 1966.

Advances have also been made in the production of apparatus, some highly sophisticated, to produce accurate rates of cell cooling, thereby improving the survival rates of cells in store. This apparatus is now of a very high standard and is to be seen in routine use in many laboratories.

THEORETICAL

No satisfactory theory which can fully explain the mechanisms of the preservation of cells has been put forward. The various methods of the preparation of cells for freezing and thawing each produce their own particular problems.

The combined studies of the structure of the cell, function and chemical composition all provide information, which when related, goes some way to providing part of an answer, but in no way the complete theory. A number of papers by established authorities in the field of cryobiology give excellent reviews on this, and the many other controversial aspects of the subject. The more important of these are, Luyet and Gehenio[16], Smith and Wood[17, 18]. Many techniques have also been developed to produce experimental results, Trump[19] and Waravdekar[20].

The effects of freezing and thawing upon the cytoplasmic elements of animal and also plant cells are well reviewed by Trump et al.[21]. They report observations made following electron microscopy, finding that the various techniques of freezing and thawing produce a wide variety of changes, with each method often having a different effect. From their work it is confirmed that the minimal cell damage is caused after very slow freezing and very rapid thawing.

When viable cells or tissues are cooled in a storage medium until freezing takes place, a number of stresses are set up within the cell. This problem is very well set out by Merryman[22]. Much of the early work on this subject has now been greatly assisted by the availability of the modern apparatus used in the field of cyrobiology.

When the temperature of a cell suspension is reduced the formation of ice crystals will commence at a temperature point below 0°C. The crystals grow as the temperature is lowered and are composed mainly of water.

Cell constituents such as electrolytes are slowly concentrated in the small amount of water remaining. The formation of ice crystals within the cell does not appear to take place unless reduction in the temperature is rapid[23]. It has been shown by Lovelock[24] that the greatest amount of damage to cells during the process of freezing is caused by the high concentrations of substances which are produced in the minute amount of water remaining unfrozen. Further work by Merryman[25] has shown that the gross cell damage caused by electrolyte concentration occurs between −5°C and −10°C. The damage is less below a temperature of −10°C.

If this damage is to be reduced to a minimum the rate of cooling of the cells from 0°C to −20°C should be slow and controlled. This will allow the formation of extracellular ice crystals and dehydration of the cells without injury. A fall of approximately one degree per minute from 0°C to at least −20°C is optimal. The upper limits of the temperature for cell storage has not been defined, but the lower limit should be below −130°C. At temperatures lower than this all chemical and physical activity are at a minimum.

A number of refrigerant liquids fall into the category of those providing a satisfactory minus zero level. The most convenient and widely used of these refrigerants is liquid nitrogen, having a boiling point of $-196°C$. No chemical nor physical changes take place at this temperature.

Liquid nitrogen has a number of advantages over other refrigerants. It is inexpensive and easily available, having no effect upon the pH of specimens and leaves no deposit upon vaporisation.

THE EFFECT OF FREEZING ON BIOLOGICAL SYSTEMS

When cells are preserved at sub-zero temperatures the aim should be to enable them to survive treatment under conditions that will cause minimal structural damage. It is difficult to judge effects because cells apparently damaged as evidenced by gross distortion, crination etc. will often survive storage, whereas cells which appear in good condition often do not survive.

These criteria do not hold good when applied to the storage of other biological materials, this must be due to the complex structure of the cell. When cells are cooled to temperatures below $0°C$, as has been described in the other sections of this chapter, three main changes occur:

1. Ice crystals are formed.
2. Water is removed.
3. The concentration of soluble materials in the cell rises.

These changes often occur at the same time.

Rapid chilling can produce a thermal shock, causing great injury to the cells and their consequent death. Lovelock[26] suggests damage to the cell membranes as a reason for this effect, although the actual cause is by no means certain. It also appears that certain cells are more sensitive to this type of injury than others[27].

The importance of considering cells as individuals becomes even more necessary when the consequences of the ice-crystal formation are considered.

If cells are cooled sufficiently slowly they will become dehydrated but not freeze intracellularly. More rapid cooling, however, will produce less dehydration of the cell, but intracellular freezing. An increase in the rate of cooling will produce intracellular ice but with a smaller crystal size. Rapid thawing is required to prevent the conversion of these small crystals to larger crystals. This would occur if thawing was allowed to progress slowly. The recrystallisation of the ice-crystals, as this phenomenon is called, is well reported[28-30]. Further observations on this aspect of the changes found in cells undergoing freezing is to be found in work by Merryman[31].

A possible explanation of the relationship between the amount of cell injury and the ice-crystal size is to be found in a paper by Mazur[32].

METHODS OF FREEZING AND THAWING

A very large volume of information has been published on these aspects of sub-zero cell storage. Some of the views will be found to conflict, but the few rules for the techniques of freezing living cells can be summarised as follows.

FREEZING

Freezing should be slow until a temperature of at least −20°C is reached. The optimum rate of reduction of temperature should be approximately 1°C per minute. The reduction in the temperature from −20°C to the storage temperature (liquid nitrogen −196°C) should then be as rapid as possible. The quantities of cells to be preserved should be small so that heat transfer through the specimens can be as rapid as possible and at the same time even.

The suspending medium should be the growth medium of the cells containing a higher than normal concentration of serum, and with the addition also of either glycerol or dimethyl sulphoxide. Reliable storage apparatus should be used. Gross cell damage can be caused in apparatus which allows a rise and fall in the basic temperature. This problem did occur when the older type solid carbon dioxide cabinets were in use, but has been eliminated by the use of liquid nitrogen cell storage containers.

METHOD

Several methods for obtaining the desired and controlled reduction in temperature are available.

The simple 'home-made' cooling plug device described by Naginton and Greaves[33] is an example of the use of the cooling effect of gas loss from a liquid nitrogen storage vessel. A development on similar lines is the Linde Cooling Plug available from Union Carbide. Both of these plug mechanisms require calibration, but can produce satisfactory results when used with care.

At the other extreme several electronic devices are to be found. One of these, first described by Pegg and Kemp, also Pegg and Trotman[34], was used for the sub-zero storage of bone marrow cells. It has been found that this type of apparatus is also suitable for cooling other cells and biological material where an accurate and programmed reduction in temperature is required. A young culture of cells is selected for storage. When in the form of a monolayer, the cells are removed from the glass surface with trypsin, washed and re-suspended in storage medium. The desired number of cells are then sealed into ampoules. For details of this method the reader is referred to Wasley and May[35].

THAWING

This process should be as rapid as possible. Ampoules should be taken from liquid nitrogen and immediately plunged into a water bath at 40°C. The cell suspension being diluted with a pre-warmed growth medium as soon as thawing is complete. This is best carried out in a covered bath. A stainless steel beaker covered with wire gauze is quite satisfactory.

STORAGE IN THE GAS PHASE

Some workers have found liquid nitrogen itself to be a source of danger in the use of liquid nitrogen refrigerators. The refrigerant can cause tissue damage on contact, and badly sealed ampoules, which may burst while thawing, can be a hazard. In the rather limited vessel neck, manipulation of ampoule holders can also prove difficult. For these and other reasons storage in the gas phase above the liquid nitrogen has now become more popular. In this method of cell preservation, the holding level for containers is raised above the level of the liquid nitrogen. As long as the refrigerant is present, the temperature will remain at a maximum of $-130°C$ and usually lower. If the vessels used for vapour storage are those specially designed for the purpose, the retrieval of the specimens is simplified and any risks are greatly reduced. Some work on techniques for holding erythrocytes in the gas phase above the liquid nitrogen level, with the cells held on plates of stainless steel gauze, has been carried out by Branson and Ginniss[36]. Transit of materials maintained at sub-zero temperatures is now possible in a vessel designed to use the gas phase produced over a plug previously soaked in liquid nitrogen.

APPARATUS SUPPLIERS

Cryogenic Equipment Sales Division,
Union Carbide Ltd.,
London, W.1.

'Linde' liquid nitrogen storage vessels
Liquid nitrogen transport vessels and trolley
Liquid nitrogen storage containers
Controlled cooling apparatus
Simple 'Linde' cooling plug

B.O.C. Cryoproducts Ltd.,
British Oxygen Company,
London, S.W.19.

'Vivostat' liquid nitrogen storage vessels
Liquid nitrogen transport vessels
Level gauges
Transfer pumps
Gas container trolleys

Messrs. Johnson and Jorgenson Ltd.,
Heringham Road,
London, S.E.7.

Glass ampoules with a low coefficient of expansion and specially suitable for sub-zero conditions.

Messrs. Scientific Supplies Ltd.,
Scientific House,
Vine Hill, London, E.C.1.

Protective face visors
Asbestos gloves

Messrs. Arnold Veterinary Ltd.,
Cremyll Road,
Richfield Estate,
Reading, Berks.

Metal clips (canes) to hold ampoules in storage canisters. Also card tube protectors to cover canes.

Messrs. Sterilin Ltd.,
Richmond, Surrey.

Plastic containers suitable for holding in liquid nitrogen.

Messrs. Baird and Tatlock,
The Matburn Agents.

Controlled cooling apparatus, programmers etc.

Spembly Technical Products,
The Trinity Trading Estate,
Sittingbourne, Kent.

A variety of items of cryogenic equipment, storage vessels, transfer vessels etc.

Chandos Products (Scientific),
New Mills,
Stockport, Cheshire.

Low temperature cabinets. Several models giving a range of temperatures from -50 to $-80°C$.

REAGENTS

British Drug Houses (Chemical Division), Poole, Dorset.	Glycerol (A.R. grade) Dimethyl sulphoxide
Messrs. Wellcome Reagents Ltd., Beckenham, Kent.	Cell Culture Media, as concentrates and in the dehydrated form. Calf serum.
Tissue Culture Services, Slough, Bucks.	Calf and foetal calf serum.
The British Oxygen Company, London, S.W.19.	Suppliers of liquid nitrogen.

REFERENCES

1. SPALLANZI (1776) cited by MANN, T., in *The Biochemistry of Semen,* Methuen, London (1954)
2. MANTEGAZZA, R. C., *1st Lombardo,* **3**, 183 (1866)
3. SHETTLES, L. B., *Am. J. Physiol,* **128**, 408 (1940)
4. PARKES, A. S., *Br. med. J.,* **2**, 212 (1945)
5. SHERMAN, J. K., *Fert. Steril.,* **15**, 485 (1964)
6. POLGE, C., SMITH, A. U. and PARKES, A. S., *Nature, Lond.,* **164**, 665 (1949)
7. LUYET, B. J., *Freezing/Drying Symposium,* p. 77. Ed. HARRIS, R. J. C., Institute of Biology
8. LOVELOCK, J. E., *Biochim. biophys. Acta,* **11**, 28 (1953)
9. MERRYMAN, H. T., *Science, N.Y.,* **124**, 515 (1956)
10. MERRYMAN, H. T., *Proc. R. Soc.,* **147**, 452 (1957)
11. REY, L. R., *Annls Nutr. Aliment.,* **11**, 103 (1957)
12. REY, L. R., *Proc. R. Soc.,* **147**, 460 (1957)
13. SMITH, A. U., *The Biological Effects of Freezing and Supercooling,* Edward Arnold, London (1961)
14. FARRANT, J., *Lab. Pract.,* **15(4)**, 402 and 409 (1966)
15. NASH, T., *Cryobiology,* Ch. 5. Ed. MERRYMAN, H. T., Academic Press, London (1966)
16. LUYET, B. and GEHENIO, P. M., 'Life and Death at Low Temperatures', *Biodynamica* (1949)
17. SMITH, A. U., *The Biological Effects of Freezing and Supercooling,* Edward Arnold, London (1961)
18. WOOD, T. H., *Adv. biol. med. Phys.,* **4**, 119 (1956)
19. TRUMP, B. F., GOLDLATT, P. J. and GRIFFIN, V. S., *Lab Invest.,* **13**, 967 (1964)
29. WARAVDEKAR, V. S., GOLDBLATT, P. J. and TRUMP, B. F., *J. Histochem. Cytochem.,* **12**, 498 (1964)
21. TRUMP, B. F., YOUNG, D. E., ARNOLD, E. and STOWELL, R., *Fedn Proc. Fedn Am. Socs exp. Biol.,* **24**, Part III, No. 2 (1965)
22. MERRYMAN, H. T., *Science, N.Y.,* **124**, 515 (1956)
23. TOKIO, N., *Recent Research in Freezing and Drying,* Blackwell, Oxford (1960)
24. LOVELOCK, J. E., *Biochim. biophys. Acta,* **10**, 414 (1953)
25. MERRYMAN, H. T., *Recent Research in Freezing and Drying,* Blackwell, Oxford (1960)
26. LOVELOCK, J. E., *Nature, Lond.,* **173**, 659 (1954)
27. SHERMAN, J. K., *Fert. Steril.,* **14**, 49 (1963)
28. MENZ, L. J. and LUYET, B. J., *Biodynamica,* **8**, 261 (1961)
29. RAPATZ, G. and LUYET, B. J., *Biodynamica,* **8**, 295 (1961)
30. REY, L. R., *Proc. R. Soc.,* **147**, 460 (1957)
31. MERRYMAN, H. T. (Ed.), *Low Temperature Research In Biology,* Academic Press, London (1966)

L

32. MAZUR, P., in *Low Temperature Research In Biology*, Ed. MERRYMAN, Academic Press, London (1966)
33. NAGGINTON, J., and GREAVES, R. I. N., *Nature, Lond.*, **194**, 4832 (1962)
34. PEGG, A. and TROTMAN, C. E., *J. clin. Path.*, **12**, 477 (1959)
35. WASLEY, G. and MAY, J. W., *Animal Cell Culture Methods,* Blackwell, Oxford (1970)
36. BRANSON, W. R. and MCGINNIS, M. H., *Blood,* **20**, 478 (1962)

SEVEN

Culture and preparation of human fibroblasts for chromosome studies

JOSEPHINE M. HYMAN and R. H. POULDING

Department of Pathology, Southmead Hospital, Bristol

INTRODUCTION

The analysis of the human karyotype from cells from both haemopoietic and connective tissues is an essential diagnostic procedure for the full cytogenetic assessment of patients clinically suspected of possessing a chromosomal abnormality. In contrast to the relatively simple two- or three-day culture of lymphocytes from blood (the commonest source of cells for chromosome studies), the culture of fibroblasts is a more complicated procedure involving a continuous culture period of several weeks before an adequate cell population is available for study. This lengthy culture period combined with a certain capriciousness of growth, varying from patient to patient, has tended to confine the culture of fibroblasts to the larger cytogenetic or research laboratories, where perhaps time is not at such a premium as the smaller hospital laboratory.

The methods available for obtaining cell concentrates of dividing human cells at the metaphase fall into three main categories depending on the type of cell under investigation. The first and most rapid are the short-term methods employing bone marrow from the sternum or iliac crest from which preparations can be made in less than 24 hours without the stimulation of a mitogenic agent[1, 2]: the second includes techniques for the proliferation of blast-like cells from mature lymphocytes in blood, spleen, lymph nodes, etc., in 48–72 hours with the addition of phytohaemagglutinin to stimulate mitotic division[3]; and thirdly, the long-term culture of fibroblasts from solid tissue such as skin without a mitogenic stimulant. Methods for the growing of fibroblasts or fibroblast-like cells can further be subdivided—those which require an initial cell suspension of fibroblasts derived by mechanical means or trypsination from the source material to initiate primary cultures[4], and those which depend on a preliminary clot embedding technique before obtaining a primary growth[5]. The following programme for the production of metaphase plates from human connective tissue has been employed routinely in a hospital pathology department for a number of years and is based largely on methods previously described[6, 7] with some modifications.

GENERAL LABORATORY REQUIREMENTS AND APPARATUS

As for any long-term culture work, the maintenance of fibroblast cultures by a series of subcultures for several weeks or longer depends on the prevention of contamination by micro-organisms by providing optimum conditions for aseptic technique. Fibroblast work can be performed in a laboratory used continuously for other laboratory work but the failure rate due to infection is high. Although antibiotics are incorporated in the basal growth medium, these provide only limited protection against accidentally introduced bacteria. A separate laboratory with washable bench tops and walls, and provided with a simple bacteriological hood, is the basic laboratory requirement if a successful fibroblast culture programme is to be maintained. It is important to remember that much of the material received for chromosome studies is irreplaceable and the remainder can only be repeated at great inconvenience to the patient. For this reason alone suitable laboratory space should be provided for those contemplating cytogenetic studies on fibroblasts.

The essential major items of laboratory equipment for fibroblast culture include incubator, waterbath, egg incubator, refrigerator, two special gas cylinders, inverted microscope and centrifuges.

37°C INCUBATOR

It is essential that the incubator selected maintain a temperature of 37°C, $\pm 0.5°C$ throughout all levels of the interior with an efficient thermostatic control. (The LEEC Precision Incubator* provides a stable temperature with minimal fluctuation and recent models include an automatic cut-out to prevent overheating in the event of thermostat failure.)

37°C WATERBATH

A standard thermostatic waterbath of internal dimensions $13 \times 10 \times 6$ in is adequate for heating media and thawing frozen reagents.

EGG INCUBATOR

For the preparation of chick embryo extract a small egg incubator or brooder is necessary to provide a regular supply of ten-day chick embryos. The Curfew electric observation incubator† is a relatively inexpensive model holding up to

* Laboratory and Electrical Engineering Company, Colwick Estates, Nottingham.
† Curfew Appliances Ltd, Ottershaw, Chertsey, Surrey.

60 eggs and requires little maintenance other than adding water to a tray to provide the correct humidity.

REFRIGERATOR

A small domestic refrigerator with an efficient freezing compartment will provide adequate storage space for reagents, media and reserve tissue.

SPECIAL GAS CYLINDERS

A supply of 5% CO_2 in air from a cylinder fitted with a fine control valve and gauge is required, preferably piped to a point under the bacteriological hood on the culture bench for convenience. A reserve cylinder will be necessary to allow for replenishment of the main cylinder.

INVERTED MICROSCOPE

For the periodic checking of cell growth in the culture vessels, an inverted microscope is used to examine the cell layer from below to avoid rotating the culture. The culture vessel rests on an adjustable stage above a nosepiece carrying the microscopic objectives with the light source and condenser above the vessel. The Prior Mark II Inverted Microscope* has proved satisfactory for the examination of cell layers in a wide range of culture vessels from the narrow Carrel flask to the deep 70 mm hexagonal baby's feeding bottle.

CENTRIFUGE

Two bench centrifuges—one with a six 20 ml bucket head and the other with a four 60 ml bucket head—will cover most requirements.

PREPARATION OF GLASSWARE

All glassware employed in fibroblast culture work requires a special washing-up procedure which must be separate from the cleaning of other laboratory glassware, particularly where contamination by siliconising fluids and strong detergents is likely. Greasy glass surfaces, or those contaminated with toxic substances, will prevent the attachment of growing fibroblasts to form a monolayer. Inadequately cleaned glassware is probably the main cause of fibroblasts failing to proliferate initially or to subculture successfully. The establishment of a rigid routine for the cleansing of glassware for fibroblast culture cannot be overemphasised.

* W. R. Prior and Co. Ltd., Bishop's Stortford, Herts.

Carrel flasks, 45×10 mm
Pyrex baby's feeding bottles (hexagonal), 125 ml capacity
Medical flat bottles, 100 ml capacity
Universal narrow-necked containers
Bijou bottles
Conical centrifuge tubes
Petri dishes, 6 cm diameter
Pasteur pipettes
Graduated pipettes

CLEANSING PROCEDURE

Solutions

Sodium metasilicate concentrate ($360 + 40$ g Calgon dissolved in 4 l water).
N/1 HCl
Standard Stergene (Lever Industries Ltd.)*

Method

Soak new and used glassware overnight in metasilicate solution diluted 1 in 100 in water, rinse in tap-water and neutralise in N/100 HCl for 1 hour. Immerse treated glassware in hot detergent—10 ml Stergene, 5 l hot tap-water. The inside surfaces are scrubbed with bottle brushes at this stage, and pipettes are washed by sucking the hot detergent solution through them several times. Rinse three times in running hot tap-water, followed by three rinses in separate changes of distilled water with a final rinse in deionised water. Bottle caps, silicone rubber liners and bungs are treated similarly. After drying, all apparatus is sterilised appropriately.

SOLUTIONS AND GROWTH MEDIA COMPONENTS

STERILE SOLUTIONS

Alsever's solution

Sodium citrate	8 g
Sodium chloride	4·2 g
Glucose	20·5 g
10% citric acid	8 ml
Deionised water to	1 l

* Lever Industries Ltd., Green Bank, London, E.1.

Sterilise by autoclaving. Distribute part of stock in 3 ml amounts in Universal containers.

Hanks' balanced salt solution

Prepare from Oxoid brand* sachets with the addition of 1·4% w/v $NaHCO_3$ adjusted to pH 7·0 with CO_2 using phenol red as indicator.

Trypsin solution

Trypsin (Difco 1 : 250)† 1 g
Hanks' solution 100 ml

Sterilise by filtration, first removing toxic impurities from the filter pad by passing deionised water through the system and discarding. For use, dilute 1 ml of stock trypsin with 20 ml Hanks' solution.

Colchicine solution

Colchicine $C_{22}H_{25}NO_6$ (B.D.H.)‡ 1 mg/ml

For use, dilute to 0·001 mg/ml with Hanks' solution. The stock solution should be prepared fresh monthly and the diluted working solution made on the day of use. If the stock solution is to be kept for more than a week, sterilise by filtration, but a small amount, e.g. 5 ml, prepared weekly and stored in the refrigerator, need not be sterilised.

Human plasma for plasma clots

Add 5 ml of fresh whole blood to 3 ml of Alsever's solution, mix by careful rotation and spin at 1000 rev/min for 10 min until the plasma/anticoagulant mixture is clear. Remove the plasma, avoiding the platelet layer, and distribute in 1 ml amounts in bijou bottles. Use fresh if possible but it will make satisfactory clots for up to a week if kept at 4°C.

To initiate the clotting mechanism, add 0·15 ml of 2% calcium chloride in deionised water/1 ml of plasma mixture.

GROWTH MEDIA COMPONENTS

TISSUE CULTURE BASAL MEDIUM

TC 199 (Morgan, Morton and Parker's medium No 199, 1950) incorporat-

* Oxoid Division of Oxo Ltd., Southwark Bridge Road, London, S.E.1.
† Baird & Tatlock (London) Ltd., Freshwater Road, Chadwell Heath, Essex.
‡ B.D.H. Chemicals Ltd., Poole, Dorset.

ing penicillin 200 units/ml and streptomycin 100 mcg/ml (wellcome Reagents Ltd.)*

CHICK EMBRYO EXTRACT (CEE)

This is best prepared in the laboratory from 10-day-old chick embryos rather than obtained commercially and the method given by Paul[8] has proved both simple and reliable. A source of good-quality fertile eggs is essential—some batches give a cloudy extract which cannot be removed by filtration at the time of preparation and which, when incorporated into the growth medium, produces a film over glass surfaces inimical to fibroblast attachment. Store CEE in 20 ml amounts in the deep freeze and, before use, thaw and spin hard to remove the fine precipitate which increases with the length of time in store. Stock CEE can be used for up to 6 weeks after preparation although the most satisfactory results are obtained from an extract less than a month old.

SERUM

Single-donor AB serum, sterilised by filtration. Batches containing obvious macroscopic lipid giving an opalescent serum should not be used. Toxic batches producing abnormalities of both nucleus and cytoplasm with chromosomal changes should be excluded by controlling each new source of serum against known actively growing fibroblasts. Pooled AB serum from several donors is often recommended to minimise the effects of a toxic batch but these sera from multiple donors can still seriously interfere with fibroblast growth if they contain a highly toxic batch. Selective control of each batch of serum whether from a single donor or pooled source is a vital safeguard. Sera from other ABO groups are satisfactory and can be used if AB serum is not available. Similarly animal sera such as calf or pig can be employed as a temporary measure if human serum is not available but the growth rate is much reduced.

GROWTH MEDIUM FOR PRIMARY CULTURES AND SUBCULTURES

MEDIUM A

Used for all primary cultures.

TC 199 with antibiotics	12 ml
AB serum	4 ml
CEE	4 ml

MEDIUM B

Used for subcultures which are required for chromosome preparations.

* Wellcome Reagents Ltd., Wellcome Research Laboratories, Beckenham, BR3 3BS.

TC 199 with antibiotics	14 ml
AB serum	4 ml
CEE	2 ml

MEDIUM C

Essentially a maintenance medium for use with stock and control subcultures.

| TC 199 with antibiotics | 16 ml |
| AB serum | 4 ml |

SKIN AND OTHER SOURCES OF FIBROBLASTS

SKIN BIOPSY

The small amount of skin required to initiate fibroblast cultures can be taken with a minimum of discomfort to the patient from suitable sites from inside forearm in adults and in the case of babies from inside thigh. Some workers prefer to use a local anaesthetic such as 1% procaine[9] but the injection of a toxic substance near the cell source is best avoided. The method described by Edwards[10] is relatively painless in skilled hands, leaves no permanent scar, and bleeding is minimal.

The selected area of skin is cleaned with 70 % spirit which is allowed to evaporate. An ellipse of skin is grasped between the shafts of a sterile pair of tapered non-toothed forceps ($4\frac{1}{2}$ in) held flat against the skin in such a way that a ridge of epidermis, about 6–8 mm long and 1–2 mm wide, is raised to a height of about 1 mm above the edge of the forceps. This is held firmly for about 30 seconds until the skin becomes white and ischaemic. A sterile surgical blade (Swann-Morton Major) is drawn smoothly along the forceps to remove the biopsy, which is transferred to a bijou bottle containing TC 199. On releasing the forceps a small amount of bleeding indicates that the biopsy is deep enough to include the dermal layer. The small wound is covered with an adhesive first aid dressing which may be removed in a few days.

OTHER SOURCES

Intersex patients with sex chromosome constitution anomalies are frequently investigated surgically to determine the morphology of primitive gonads. Biopsies of such gonadal tissue and more fully differentiated ovaries or testes can be used as a source of fibroblast material. Fragments should be collected in TC 199 and not come in contact with formalin or any fixative which may be used to fix similar biopsies for histological studies. Needle biopsies or fragments taken by open surgery from testes in cases of Klinefelter's syndrome (XXY) are a common source of fibroblasts and a fragment of the main biopsy taken for histology is all that is required for fibroblast culture.

Foetal material from surgical abortion is an excellent source of fibroblasts and is of special value when the foetus is suspected of being abnormal, particularly as fibroblasts are usually the only method of determining the chromosome constitution. Spontaneous abortions are invariably contaminated unless the foetal sac is intact as many have died some time before they are aborted. Such material should be avoided. Skin taken *post mortem* from stillbirths (non-macerated), neonates and older cadavers can provide successful material providing the body has been refrigerated and is less than 24 hours *post mortem*. Biopsies from cadavers should be stored at least one day in TC 199 with antibiotics (changed several times) to minimise possible bacterial contamination.

MONOLAYER CELL CULTURE METHOD

SETTING UP OF PRIMARY CULTURES

Fibroblasts and fibroblast-like cells derived from skin and other tissues grow directly along the surface of the culture vessel forming a monolayer firmly adherent to the glass. When the available area for cell growth is covered, the fibroblasts grow over one another to form a layer several cells thick, which if left would eventually become detached from the glass to form a 'ball' of cells. The optimum stage to arrest the culture for subculturing is before the whole surface area is utilised and while the monolayer is still actively expanding. Primary cultures are set up in Carrel flasks, 45 mm in diameter and 10 mm in depth, with the narrow neck sealed with a tight fitting silicone rubber bung. The main advantages of this type of vessel for initiating fibroblast cultures are the relatively large growth area compared with the volume of culture medium; the small volume of atmosphere above the medium, which facilitates the maintenance of an optimum pH; and the design allows for easy control of aseptic conditions over a long period.

Initial *in vitro* division requires that viable fibroblasts remain immobilised and in contact with the glass surface for a variable 'lag period' during which they become adjusted to artificial conditions. Three methods of setting up primary cultures are described.

1. DIRECT CONTACT CULTURE METHOD

Initial fibroblast growth can be obtained direct from the rapidly dividing tissues of foetal and neonatal origin by placing a number of the tissue fragments in a Carrel flask with sufficient Medium A to barely cover the base of the flask, and replacing the atmosphere with 5% CO_2 in air. The flask is left undisturbed in the incubator for 48 hours, after which it may be examined with the inverted microscope for fibroblast outgrowths. At the onset of cell growth the fragments become firmly attached to the glass, and the total volume of medium can be increased to 2 ml to allow for normal monolayer expansion.

2. TRYPSINISATION METHOD

A suspension of cells is made by finely chopping the skin and digesting the tissue matrix with a trypsin solution. The skin biopsy is placed in a sterile 6 cm Petri dish in a pool of sterile, isotonic 0·04% trypsin solution[4], and cut into small fragments with sterile instruments. After incubation at 37°C for about 30 min, with occasional agitation of the fluid by repeatedly sucking it into a Pasteur pipette, a cloudy suspension of cells results. This is centrifuged and the deposit inoculated into 1–2 ml of Medium A in a Carrel flask, which is incubated at 37°C after replacing the atmosphere with 5% CO_2 in air. Individual cells and small tissue fragments which settle on the glass surface should commence to divide after a few days. A similar technique is described by Fraccaro, Kaijser and Lindsten[11].

In this laboratory this method has proved impracticable with a low success rate. Fibroblasts not adapted to *in vitro* culture do not settle and attach themselves readily to the glass, and even slight disturbance of the cultures (for example, by convection currents within the culture medium) affects the attachment of the cells to the glass surface. However, the technique described by Hsu and Kellogg[12] (adapted from a method devised by Evans and Earle[13]) using sterile discs of perforated Cellophane to hold cell aggregates in place until cultures are established may have overcome this difficulty of cell movement.

3. PLASMA CLOT EMBEDDING TECHNIQUE

Fragments of skin (explants) are attached to the glass surface in plasma clots, and remain undisturbed, whilst the new growth of cells radiates outwards. The skin biopsy is placed in a sterile 6 cm Petri dish in a small pool of TC 199 and cut into a dozen fragments about 0·5 to 1 mm in diameter with sterile fine forceps and scissors. A black tile or black paper beneath the Petri dish is a considerable help for this fine dissection. If these fragments are embedded in plasma clots immediately, they will often become detached from the clots by enzymic action over several days and will require re-embedding in new Carrel flasks. This proteolytic digestion of the plasma clot can be overcome by placing the fragments of the biopsy in a Carrel flask with 2 ml of Medium A, replacing the atmosphere with 5% CO_2 in air and incubating 3–7 days before embedding as described below. By using this preliminary incubating stage before embedding, the danger of premature clot detachment is almost eliminated, and the onset of growth is not delayed.

Routinely, two primary cultures are set up from one biopsy, each with about six explants. The skin fragments are transferred individually to a small pool of the prepared human plasma with anticoagulant in another sterile Petri dish, and agitated to remove traces of TC 199. The minute explants are most conveniently carried from one vessel to another by means of a Pasteur pipette which has a small constriction pulled near its tip, which prevents loss of the tissue by suction into the main body of the pipette. Most of the plasma is now removed from around the explants, which are transferred as dry as possible, and arranged in the two Carrel flasks. A further sample of the prepared human plasma is reconstituted by adding 0·15 ml of sterile 2% calcium chloride to 1 ml of plasma and mixing thoroughly. With a fine Pasteur pipette a small drop of this is superimposed over each explant, removed and replaced with a fresh drop, making sure that the tissue is completely surrounded by the plasma and not floating on its surface. The size of the drop is adjusted to give a clot capable of attaching the skin to the glass surface without being excessively large. Once the plasma has been reconstituted a period of about five minutes elapses before clotting occurs, but this varies for different donors and depends on the degree of freshness of the sample. However, with a little practice this time is more than adequate to complete the operation. The Carrel flasks with cotton wool plugs should be placed in the incubator for at least 30 min to ensure firm clot formation before 2 ml of Medium A are carefully added and the gas phase adjusted with 5% CO_2 in air. The cotton wool plugs are replaced with silicone rubber bungs.

Owing to faulty technique occasionally some, if not all, of the explants may become detached within a few hours. In such cases it is advisable to remove the loose explants, wash in fresh plasma and re-embed them in a new Carrel flask rather than to repair existing clots.

TREATMENT OF MONOLAYER TO OBTAIN METAPHASE HARVEST

SUBCULTURE

Following the emergence of the first few fibroblasts from the explant, cell duplication proceeds rapidly. After about five to ten days, depending on how many of the explants are proliferating, examination of the culture with the inverted microscope shows confluent sheets of fibroblasts covering about one-third to half the total area of the Carrel flask base, and at this critical stage subculture into larger culture flasks is performed.

The culture medium in the Carrel flask is replaced with a weak isotonic solution of trypsin, and this mild digestive process causes the fibroblasts to become spherical and to lose their attachment to the glass. After about ten minutes digestion at 37°C, the resultant cell suspension is transferred to a sterile plugged conical tube and centrifuged at 1000 rev/min for five minutes. The button of cells is resuspended in 5 ml of Medium B and poured into the subculture vessel. The gas phase is adjusted with 5% CO_2 in air, as for the primary cultures, and the subculture is incubated at 37°C. Within two to three hours the cells have settled and flattened out attaching themselves to the glass surface before proliferating. Routinely, the growth medium is replaced with fresh medium three times a week.

Pyrex baby's feeding bottles sealed with a silicone rubber bung are ideal culture vessels; a monolayer of fibroblasts on one of the flat surfaces provides a convenient number of cells for treatment to produce metaphase plates for chromosome examination. When the first subculture, or first passage of cells has proliferated sufficiently to form a monolayer, this can either be treated with an anti-mitotic agent to arrest the cells in the metaphase prior to producing chromosome preparations, or the monolayer can be retrypsinised, yielding a suspension of cells which is divided and inoculated into other feeding bottles to provide two or more subcultures at the second passage (Fig. 7.1). In this manner alternative cultures provide an unlimited supply of cells for further investigation should the initial chromosome preparations prove unsatisfactory.

The original trypsinised primary cultures will, in most cases, yield a further growth of fibroblasts if the Carrel flasks are re-incubated with fresh culture medium. This can prove invaluable in the event of failure of the subculture.

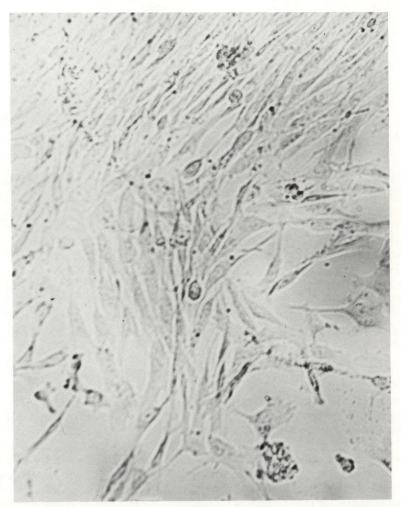

Fig. 7.1. Fibroblast monolayer in a subculture bottle viewed with the inverted microscope. Note the rounded cells amongst the attenuated fibroblasts indicating a high mitotic activity in a subculture suitable for chromosome preparations. (× c. 130).

OPTIMUM CONDITION OF FIBROBLAST MONOLAYER

The preparation of metaphase plates (Fig. 7.2) for chromosome studies depends on the termination of a monolayer culture at a time when a proportion of the cells are in mitosis. Fibroblasts in mitosis are characterised by a change in morphology from the normal attenuated spindle form to a spherical configuration readily recognised in active cultures under the inverted microscope. By this means some indication of the mitotic index can be estimated.

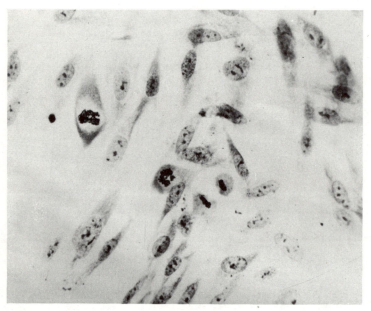

Fig. 7.2. Fibroblast monolayer showing three cells in various stages of mitosis. Buffered thionin stain. ($\times 225$)

The peak mitotic activity for early subcultures occurs in approximately 18–24 hours after the inoculation of the culture vessel with a concentration of not less than 5×10^5 cells in 5 ml of Medium B.

An increase in mitotic activity can often be obtained by treating the subcultures with cold shock at a temperature of 4°C for 30–60 min which partially synchronises the phases in the total generation time[14]. Early on the day following trypsinisation of cells for the seeding of cultures for chromosome preparations, the feeding bottles are placed in the refrigerator at 4°C, usually for 30 min. They are then replaced in the incubator at 37°C, and left to readjust to this temperature for 30 min before further treatment. It is not always necessary to use cold shock if the previous trypsinisation for passage of the cells was within 24 hours, but application of the technique usually gives an adequate yield of mitotic cells in subcultures of fairly long standing, which would not have otherwise produced satisfactory results. As a routine, cold shock is used in conjunction with recent passage of the fibroblasts to give optimum mitotic activity prior to growth termination.

COLCHICINE TREATMENT

Although satisfactory metaphases can be obtained without the use of colchicine to arrest dividing cells at the metaphase by inhibiting spindle formation, a consistently higher yield of metaphase plates is obtained when colchicine is added to the cultures from 4–6 hours prior to termination. 0·2 ml of the dilute working solution of colchicine (0·001 mg/ml in Hanks' Solution) is added for every 5 ml of medium after rewarming the culture following cold shock. After a further incubation time of about 5 hours the culture is terminated by trypsinisation to collect a suspension of separated cells. After centrifuging, the trypsin solution is discarded and the cells carefully resuspended in warmed Hanks' solution.

POST-COLCHICINE TREATMENT

In the technique described by Harnden[5] the final subculture was made into a Petri dish containing a number of glass coverslips. The fibroblasts which then grew over these were treated *in situ* to produce preparations for chromosome study. Apart from the minor technical difficulty of handling fragile coverslips, this method had the disadvantage that fibroblasts in mitosis, having become spherical in shape, showed a tendency to become detached from the glass surface, and therefore lost into the surrounding medium. The attachment of the mitotic cells remaining, resulted in inadequate spreading of the chromosomes. Frøland[15] combined Harnden's culture technique with the method of preparing metaphase plates for chromosome studies from venous blood, devised by Moorhead *et al.*[3] treating the cells as a suspension in the final stages. The following technique gives consistent success with well-spread metaphase plates from cell suspensions with a high mitotic index.

Following colchicine treatment, the supernatant medium is decanted into a graduated conical centrifuge tube and spun at *c.* 1000 rev/min to recover any mitotic cells which have become detached from the growing surface. Meanwhile the monolayer is disaggregated with 0·05% trypsin at 37°C as if for subculture. When all the fibroblasts are in suspension, the fluid is added to the cell deposit in the centrifuge tube, further dispersed by gentle pipetting, and re-spun. Slight cell loss on to the glass surfaces is prevented if the conical centrifuge tube and the pipette are pretreated with siliconing fluid. Adequate separation of the cells, providing a uniform fine suspension is important, and is facilitated by the selection of a fine bore pipette with an aperture less than 1 mm in diameter.

HYPOTONIC TREATMENT

In order to ensure adequately spread metaphase plates, it is necessary to enlarge the cells sufficiently to give well spaced chromosomes in a cytoplasm which is not too dense, yet stable enough to prevent scattering of the chromosomes by rupture of the cell membrane. These criteria are achieved by exposing the cells to a hypotonic solution. Fluids which have been used for this purpose include sodium citrate, 0·95% (Frøland[15]) and 0·7% (Hirschhorn and Cooper[9]), 0·17% saline (Tjio and Puck[16]), 0·075 M potassium chloride (Hungerford[17]), and various dilutions of Hanks' buffered salt solution (e.g. Harnden[5]). Whilst all these hypotonic solutions give satisfactory metaphase expansion, the one adopted for routine use is a quarter of the strength of Hanks'. Following the washing of the cells in full strength Hanks' solution, most of the supernatant is removed, leaving the cells suspended in 0·5 ml of solution. This is then carefully diluted drop by drop with warm distilled water to give a final volume of 2 ml $\frac{1}{4}$ strength Hanks', which is allowed to act for 15–20 min at 37°C.

FIXATION AND SPREADING TECHNIQUE

The button of cells in the conical tube following centrifugation and removal of the supernatant hypotonic solution is fixed undisturbed for thirty minutes in a mixture of methyl alcohol and glacial acetic acid (3:1); best results are achieved if the fixative is freshly prepared immediately before use. After this time the supernatant is removed and the cells resuspended in fresh fixative. At this stage, optimum success depends on repeated pipetting of the fixed cells with the siliconed Pasteur pipette, to ensure a fine, even cell suspension. Following centrifugation the cells are washed in this way in a total of three changes of freshly prepared acetic alcohol; inadequate spreading may necessitate further changes. Finally, the cells are suspended in a small volume of fixative to give a milky suspension which is just opaque, but not dense. Normally the total volume required is between 0·25 and 0·5 ml.

Spreading of the mitotic cells is carried out by a modification of the air-drying method described by Rothfels and Simonovitch[18]. Alcohol cleaned, grease free slides are chilled in a Coplin jar of distilled water until ice crystals have begun to form. On withdrawal of a slide, it is covered by an even film of ice-cold water. Holding the slide at an angle of about 45°, a drop of cell suspension is allowed to run from the top; gentle blowing helps to speed up the spreading of the cells over the total surface area. The preparation is dried by vigorously waving the slide in the air, and final drying is accomplished over the flame of a spirit lamp. The drying process must be rapid and should be accomplished within 30 seconds after which the slide is examined microscopically to assess the degree of spreading of the mitotic cells. With practice consistently good results have been achieved. Failure to obtain well-spread metaphase plates can sometimes be rectified by 'flaming' the slides, that is by igniting the fixative as soon as the drop of cell suspension has been allowed to run over the surface of the wet, chilled slide, but this more drastic technique may result in artefacts, such as fractures of the chromatids, and loss of chromosomes due to over-spreading and rupture of the cell membrane. It is therefore advisable to avoid the 'flaming' technique where possible.

Slides made by the air-drying technique are permanent and can be stored indefinitely either prior to staining or as stained, mounted preparations. This is a distinct advantage over the original 'squash' techniques which are more difficult to make into permanent preparations and require more skill to give consistent results of the high standard achieved by the air-drying technique.

After spreading, the slides are stored in methyl alcohol, as this secondary fixation in the absence of acetic acid enhances the subsequent staining procedure.

STAINING OF FIBROBLAST METAPHASE CHROMOSOMES

The three basic methods used for staining human chromosomes in leucocyte material—Feulgen, aceto-orcein variants and Romanowsky-type stains—can be employed to stain chromosomes in fibroblast preparations. However, when routine bulk staining is necessary a Romanowsky stain is preferred as this gives a more consistent depth of colour to the chromosomes which stain a deep violet to purple. As with both Feulgen and aceto-orcein stain there is a tendency for Romanowsky stain to fade unless stored in the dark and they are cleared and mounted from high quality, sulphur-free xylene. The following May-Grünwald-Giemsa variant is both simple and reliable for the routine staining of air-dried preparations in bulk.

REAGENTS

M/10 Sørensen's phosphate buffer mixture at pH 6·8 diluted with water 1:20.

Filtered May-Grünwald staining solution—0·3 g May-Grünwald powder (G. T. Gurr)* in 100 ml methyl alcohol dissolved gradually at 37°C in waterbath. For use, dilute 2 parts with 3 parts 6·8 buffer.

G. T. Gurr's improved Giemsa stain R66. For use, dilute 1 part with 4 parts 6·8 buffer.

METHOD

1. Rewash smears in fresh methyl alcohol.
2. Stain in freshly diluted May-Grünwald stain 3–5 min.
3. Drain off excess stain and replace with diluted Giemsa, 20 min.
4. Wash briefly with running tap water adjusting the direction of the flow to prevent the surface scum from settling on the cells.
5. Differentiate in 6·8 buffer until cytoplasm surrounding chromosomes is clear. In most cases a brief wash in buffer suffices.

* G. T. Gurr Ltd., Carlisle Road, The Hyde, London, N.W.9.

M

6. Blot gently with fluffless filter paper, removing most of buffer and finally air dry.
7. Clear in two changes of new xylene and mount in D.P.X.
8. Store away from light.

ANALYSIS AND PHOTOGRAPHY OF FIBROBLAST METAPHASES

The visual analysis by direct microscopy of spread metaphases from fibroblasts is similar to the standard procedure used for transformed lymphocyte metaphases obtained from cultured blood leucocytes. The chromosomes of fibroblasts in metaphases prepared by the above method are frequently less

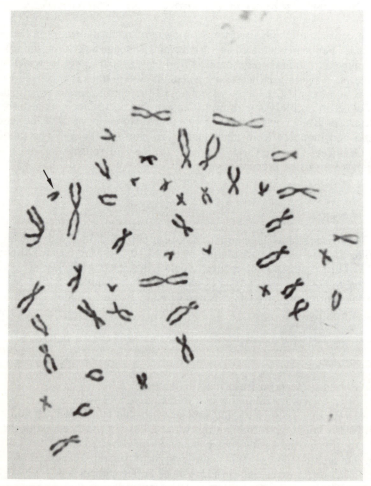

Fig. 7.3. Photomicrograph of fibroblast metaphase plate from a normal male, 46,XY. The Y chromosome is arrowed. May-Grünwald–Giemsa stain (× c. 1000)

contracted than those made from leucocyte cultures and as a result are thinner and paler stained but often show greater detail in areas of secondary constrictions such as in the satellited stalks on the D and G group chromosomes. The classification and identification of the individual chromosomes follows the standard system recommended by the Denver Study Group (1960)[19] and the

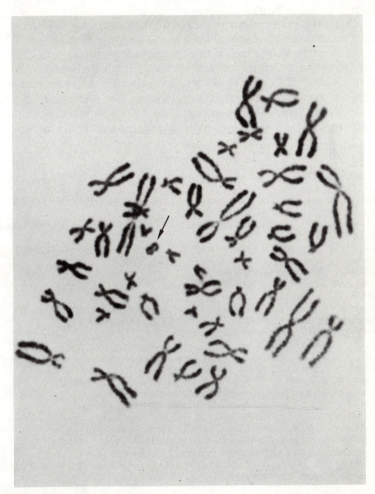

Fig. 7.4. Photomicrograph of fibroblast metaphase plate from a patient with an abnormal chromosome—one of pair Group E 17 is in the form of a ring chromosome (arrowed). May-Grünwald–Giemsa stain. (× c. 1000)

London Conference (1963)[20] whilst the standard nomenclature adopted by the Chicago Conference (1966)[21] can be used to describe the many numerical and structural abnormalities of the human chromosome complement.

Photomicrography of metaphase plates for record purposes and for karyotyping abnormal chromosome complements is an essential adjunct to chromosome analysis (Fig. 7.3 and Fig. 7.4). To obtain maximum resolution and contrast of the reddish-purple chromosomes stained by May-Grünwald-Giemsa, it is necessary to use high resolution, maximum contrast photographic film. For 35 mm cameras Kodak Recordak Micro-file film (Type 5669) developed in either Kodak D8 maximum contrast developer or May and Baker Qualitol gives excellent negatives of suitable contrast without loss of the finer structures such as satellites when used in conjunction with a green filter, e.g. Wratten 58. For use in plate cameras Ilford chromatic G30 plates or Kodak Commercial Ortho sheet film are recommended followed by development in Qualitol diluted 1:19. Prints, approximately 12×16 cm, are large enough for recording a complete metaphase but where karyotyping is required, prints should be larger, e.g. 16×21 cm to allow the individual chromosomes to be easily handled for arranging in a karyotype.

REFERENCES

1. FORD, C. E., JACOBS, P. A. and LAJTHA, L. G., *Nature, Lond.,* **181**, 1565 (1958)
2. TJIO, J. H. and WHANG, J., *Stain Technol.,* **37**, 17 (1962)
3. MOORHEAD, P. S., NOWELL, P. C., MELLMAN, W. J., BATTIPS, D. M. and HUNGERFORD, D. A., *Expl. Cell Res.,* **20**, 613 (1960)
4. PUCK, T. T., CIECIURA, S. J. and ROBINSON, A., *J. exp. Med.,* **108**, 945 (1958)
5. HARNDEN, D. G., *Br. J. exp. Path.,* **41**, 31 (1960)
6. HYMAN, J. M., *J. med. Lab. Technol.,* **25**, 81 (1968)
7. POULDING, R. H., *Progress in Medical Laboratory Technique*, vol. 4, Butterworth, London (1968)
8. PAUL, J., *Cell and Tissue Culture*, 2nd edn, Livingstone, Edinburgh (1960)
9. HIRSCHHORN, K. and COOPER, H., *Am. J. Med.,* **31**, 442 (1961)
10. EDWARDS. J. H., *Lancet,* **i**, 496 (1960)
11. FRACCARO, M., KAIJSER, K. and LINDSTEN, J., *Ann. hum. Genet.,* **24**, 45 (1960)
12. HSU, T. C. and KELLOGG, D. S., *J. natn Cancer Inst.,* **25**, 221 (1960)
13. EVANS, V. J. and EARLE, W. R., *J. natn Cancer Inst.,* **8**, 103 (1947)
14. SINCLAIR, W. K. and MORTON, R. A., *Nature, Lond.,* **199**, 1158 (1963)
15. FRØLAND, A., *Acta path. microbiol. scand.,* **53**, 319 (1961)
16. TJIO, J. H. and PUCK, T. T., *J. exp. Med.,* **108**, 259 (1958)
17. HUNGERFORD, D. A., *Stain Technol.,* **40**, 333 (1965)
18. ROTHFELS, K. H. and SIMINOVITCH, L., *Stain Technol.,* **33**, 73 (1958)
19. DENVER STUDY GROUP (1960), *Lancet,* **i**, 1063 (1960)
20. LONDON CONFERENCE (1963), *Cytogenetics,* **2**, 264 (1963)
21. CHICAGO CONFERENCE (1966), *Birth defects: Original article series,* **II**, 2, The National Foundation, New York (1966)

EIGHT

The fluorescent antibody technique as applied to tissue culture monolayers

SHEILA HUDSON

Department of Clinical Virology, St Thomas's Hospital, London

INTRODUCTION

This chapter describes the basic principles of the fluorescent antibody (FA) technique, with particular reference to its application in tissue culture and to the methods employed in a diagnostic virus laboratory.

The application of the FA techniques to individual viral systems is not included, but as an introduction to the chapter, a brief summary of some of its advances in the field of rapid diagnosis will be given.

The fluorescent antibody technique was first used by Coons[1], for the demonstration of pneumococcal antigen. Since then it has been widely used for the detection of both bacterial and viral antigens. At one time it was used basically as a research tool, but now it is being increasingly employed as a method of achieving a rapid laboratory diagnosis for many diseases. The technique may be used to (1) test for the presence of antigen, using a serum of known antibody content, and (2) to detect the presence of a particular antibody using a known antigen. Throughout the chapter, reference is made only to the first of these applications.

As a diagnostic tool, the test is of particular value when antigen can be detected in material obtained directly from the patient, for example, in cells from the throat, nasal washings, or urine. Liu[2] detected Influenza A antigen in cells obtained from nasal washings. Hers[3] has also detected this antigen in smears from sputum, and Llandes-Rodas and Liu[4] diagnosed measles from epithelial cells in urinary sediments. Gardner and McQuillan[5] have observed RSV antigen in epithelial cells obtained from naso-pharyngeal secretions. More recently[6], rubella-virus antigen has been demonstrated in cells obtained from throat swabs.

It is not always possible to establish a diagnosis using exfoliated cells; too few cells may be obtained, or only a small percentage of those obtained may contain antigen. It is, therefore, advisable to inoculate some of the specimen into tissue culture. Depending on the replication of the virus and the particular cell system used, the antigen can be detected within these cells by the FA technique at various times after infection. This method was employed successfully by McQuillan and Gardner[7] who detected RSV in the early stages of replication.

Before this technique can be discussed (in relation to the detection of viral antigen in tissue culture), a description is given of the materials and methods used, and details of some of the difficulties encountered and how they may be overcome.

The two methods most commonly employed in fluorescent antibody tracing are the direct method of Coons and the indirect method of Weller and Coons[8].

1. THE DIRECT METHOD

The specific γ-globulin fraction of the serum is directly conjugated to the fluorescent dye; the product itself being referred to as the conjugate. This is applied to the tissue containing the fixed antigen and allowed to react in a humidified chamber of 37°C for 30–60 minutes. The preparation is given a preliminary wash to rinse off excess dye, it is then immersed in phosphate buffered saline (PBS) for a minimal time of 30 minutes. During this period gentle agitation and several changes of the PBS are used to completely remove any conjugate which may adhere to the tissue but is not associated with the antigen.

The preparation is mounted in a suitable medium on a clean glass slide and examined using a fluorescent microscopy.

2. THE INDIRECT METHOD

The unconjugated immune serum is first allowed to react with the tissue antigen when using this method. The preparation is then washed thoroughly to remove any uncombined antibody. The conjugated protein, which is anti-species, i.e. directed against the source of immune serum used, is now applied and again allowed to react for 30–60 minutes at 37°C. The preparation is now treated as in method 1.

COMPLEMENT STAINING

This technique is not so widely used. Goldwasser and Shephard[9] found that by using antiserum from various sources in the presence of guinea-pig complement and then applying an anti-guinea-pig conjugate, the presence of antibody could be detected in dilutions in which staining against the specific antibody itself was weak or negative. When the antiserum used was not of guinea-pig origin, it was necessary to inactivate this at 56°C for half an hour to avoid competition between the different types of complement for the antibody/antigen complex.

The advantage of this method is the convenience of using an anti-guinea-pig conjugate for antisera of any origin, together with the fact that the sensitivity of the indirect method appears to be increased provided antisera are inactivated first.

ADVANTAGES AND DISADVANTAGES

As can be seen from Figs. 8.1 and 8.2, the direct method is a one step reaction, whereas the indirect method involves two steps. The direct method is also quicker, which is an advantage in diagnostic work. It is sometimes necessary to

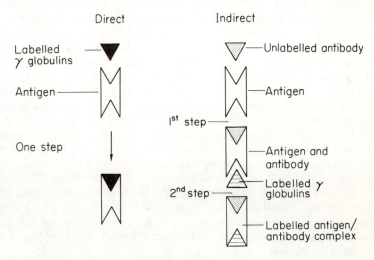

Figs. 8.1 and 8.2. Diagrammatic representation of the reactions involved in (Fig. 8.1) direct (Fig. 8.2) indirect fluorescent antibody staining

conjugate serum; this is not a difficult procedure, but a laborious one. The indirect method has the advantage that the one conjugate can be used for a variety of antigens. Conjugates are now prepared commercially by several firms but the quality may vary considerably. It is, therefore, best to test several under the conditions of one's own system to determine which is the most satisfactory.

The indirect method is generally more sensitive, it can be regarded as a double antigen/antibody reaction in which the first reaction becomes the antigen of the second step.

THE FLUORESCENT COMPOUND

A substance is said to fluoresce if it absorbs light energy of one wave length, raising it to an 'excited' state, and emits light of another wave length. Due to energy loss the emitted light is always of longer wave length than the excited light, this being independent of the wave length of the absorbed light. Fluorescein fluoresces in the yellow-green region of the spectrum regardless of whether it is illuminated by ultra-violet or blue light.

The dyes most commonly used for fluorescent microscopy are lissamine rhodamine B (RB 200), Dans (1-dimethylamino-naphthalene,5-sulphonic acid) and fluorescein isothiocyanate (FITC), the last of these superseding fluorescein isocyanate which Coons originally used, the thiocyanate being chemically more stable. Details concerning conjugates are beyond the scope of this chapter but for details of the procedure for conjugation of sera, reference should be made to Nairn[10].

For a substance to be of use as a conjugate it must be reasonably stable in its combined or conjugated form. Various factors are liable to alter the combined form, e.g. pH of medium or temperature. In general the higher the temperature the less stable the conjugate. Conjugates will keep several months without loss of activity stored in sealed aliquots at $-20°C$. (In some cases the manufacturer may recommed $+4°C$ for a particular product.) It is also essential that the antibody activity of the serum is maintained after conjugation and that the fluorescent property of the dye should decrease as little as possible. The fluorescent colour of the conjugate should be different from that of the background tissue.

All the microscopical equipment should be kept as clean as possible to minimise the accumulation of dust which may give rise to particulate fluorescence. Microscope slides should be of 1 mm thickness and manufactured from preferably fluorescence-free glass. Immersion oil should be non-fluorescent and kept in a small oil can in preference to a bottle which necessitates repeated removal of the cap and increases the possibilities of contamination by dust. The mountant must also be fluorescence free. For semi-permanent preparations, phosphate buffered glycerol pH 7·0–7·4 can be used. 'DPX' or 'Polafluer B' is used as a permanent mountant.

The object, above which is placed an ultra-violet absorption filter to protect the observer's eyes from harmful rays, is illuminated by a powerful light source of ultra-violet blue light, e.g. a high pressure mercury vapour lamp such as an HBO 200 watt bulb. The lamp is completely enclosed to avoid dazzle and ultra-violet radiation, and as a safeguard should the lamp explode. The normal life of a bulb is in the order of 200 hours, but the efficiency tends to fall noticeably after about 150 hours use. A record should be kept of the hours the lamp is used and two hours are added each time the lamp is switched on and off. A starter unit provides the high tension necessary to strike the arc. Within the lamp housing is fitted the collector lens which is usually fixed in position by the manufacturer. This is followed by a heat absorption filter which is also fixed in position. Each new lamp which is inserted must be centred. This is done by placing a piece of thin paper on the exit aperture in the base of the microscope. The field diaphragm is slightly closed and then the image of the arc between the two electrodes is adjusted until it lies at the centre of the exit aperture.

The HBO 200 lamp has an emission range of 280–600 nm with its main peaks at 365 nm and 435 nm (Fig. 8.3). For work with ultra-violet light using a dark ground condenser, the practical excitation range of FITC is between 290 nm and 400 nm (Fig. 8.4). The green emission of FITC is best seen against a

blue background of tissue autofluorescence. This is achieved by using a
secondary filter above the object which cuts off light up to 385 nm. The dark
ground condenser is especially useful in oil immersion microscopy as it reduces

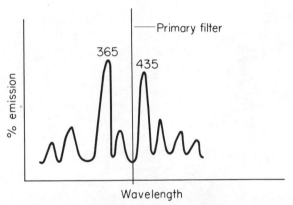

Fig. 8.3. Emission spectrum of HBO 200 mercury lamp

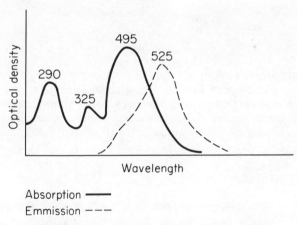

Fig. 8.4. Absorption and emission spectra of FITC

the 'bloom' in the oil by reducing the area illuminated. This condenser also
scatters the light and thus directs it away from the objective allowing a greater
proportion of the spectrum of the primary source to be used to stimulate
fluorescence; a black background can still be produced by a relatively thin
barrier filter. A primary filter should be used which transmits only ultra-violet

light and cuts out visible light and heat. Thus the only light reaching the object has a wave length of up to 400 nm (Fig. 8.5). The absorption peaks of fluorescein isothiocyanate are 290, 325 and 495, with an emission peak of 525 nm.

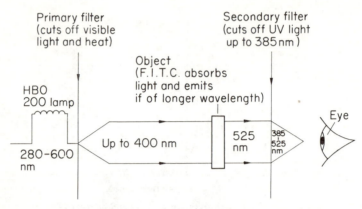

Fig. 8.5. The relationship between various components of the microscope, wavelength, absorption and emission

PHOTOGRAPHY

All photographs were taken with an EXA IA camera. For colour transparencies, Kodak Ektachrome Type B for artificial light gave very satisfactory results. The exposure times varied between $2\frac{1}{2}$–3 minutes, depending on the intensity of the fluorescence. For black and white photography Kodak Tri-X film was used with similar exposure times (Figs. 8.6–8.10).

NON-SPECIFIC STAINING

One of the most important requirements of the fluorescent antibody technique is that the staining is specific for the particular antigen/antibody system under test. It is essential that the only place where fluorescence is detectable is at the site of an antigen/antibody reaction. Where staining of the tissue is not due to such a reaction it is termed non-specific. It can be caused by free fluorescent dye in the conjugate, by conjugated serum proteins or by unwanted conjugated antibodies.

The free dye fluorescence of the conjugate can be removed either by gel filtration with Sephadex or polyacrylamide. This, however, is not usually a problem encountered when using conjugate from a commercial source unless the conjugate has been poorly stored and the dye has dissociated.

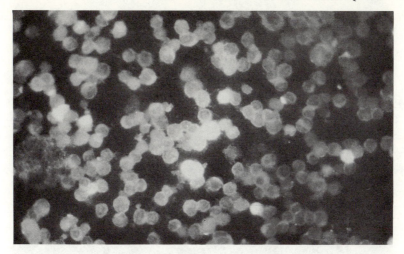

Fig. 8.6. EB$_3$ cells fluorescing

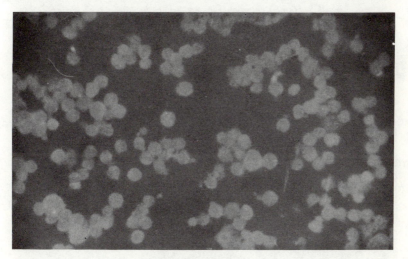

Fig. 8.7. EB$_3$ cells (control)

Nairn[11] suggests that electrostatic forces between the microscope preparation and the serum proteins are responsible for the staining produced by conjugated serum proteins. Such proteins behave as acid dyes and stain positively charged tissue proteins. They can be removed by absorbing the serum with tissue powders. Fluorescent conjugates, by the same principle, can be treated in this way.

The following method is used in the author's laboratory.

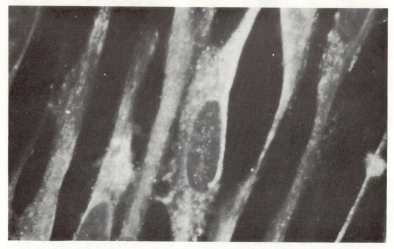

Fig. 8.8. Human embryonic lung. Chronically infected cell line

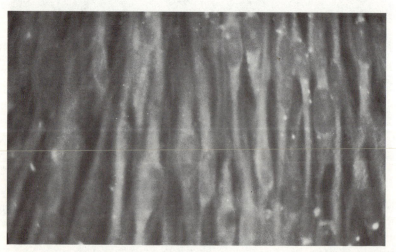

Fig 8.9. Human embryonic lung (control)

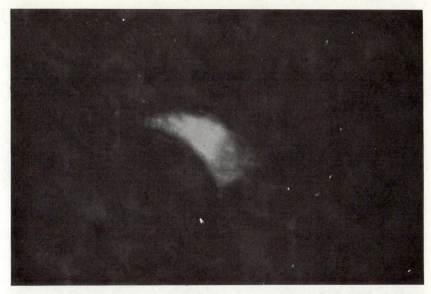

Fig. 8.10. RK$_{13}$ cell. Infected with Rubella virus. 3 days. Single cell showing cytoplasmic fluorescence

ABSORPTION OF CONJUGATE WITH LIVER POWDER

PREPARATION OF LIVER POWDER

Fresh livers from adult mice, which have been killed by dislocation of the neck (not with anaesthetics), are chopped into small fragments with scissors. Approximately 1 ml of tissue fragments is transferred to sterile tubes. A small piece of sterile hydrophilic gauze is placed on top of the tube and the tissue frozen by holding the tubes in a container of freezing mixture, i.e. dry ice and alcohol. The contents of the tubes are then freeze-dried for twenty-four hours, after which the dried tissue fragments are transferred to a sterile mortar and ground to a powder. The liver powder is subsequently stored in stoppered tubes at +4°C. Commercially, a very good preparation of liver powder, of pig origin, can be obtained from Wellcome Reagents Ltd.

ABSORPTION

The freeze-dried liver powder is washed twice with sterile distilled water and once with phosphate buffered saline pH 7·4, the suspension being centrifuged after each washing. The washed liver powder is finally added to the conjugate, in the proportion of 1 part washed powder to 5 parts conjugate, and absorbed for one hour at room temperature. The mixture is stirred, care being taken to avoid the formation of air bubbles. (A mechanical stirrer may be used at a low speed, or the mixture may be gently agitated by hand every five minutes.) This suspension is then centrifuged for 30 minutes at 10 000 rev/min at +4°C. The conjugate is carefully removed and re-absorbed for a further one hour with freshly washed liver powder and is again centrifuged. The clear supernatant is pipetted into stoppered ampoules in small aliquots and stored at −20°C.

STAINING BY UNWANTED ANTIBODIES IN THE SERUM

Often when a virus has been grown in tissue culture and then inoculated into an experimental animal to produce an antiserum to this virus, antibodies not only to the virus but also to cellular material inoculated along with it, are produced. For example, if an antiserum to RSV prepared in rabbits and an anti-rabbit conjugate were applied to normal uninfected HeLa and Hep-2 cells with the result that fluorescence was obtained, then it would be essential to remove this non-specific staining in order to avoid obtaining false positive results. It is often convenient and effective to absorb such a serum with preparations of the normal tissue being used in the system studied. A successful method for doing this is described by Sander[13].

SERUM ABSORPTION WITH NORMAL TISSUE

Each 1 ml of antiserum is absorbed with HeLa and Hep-2 cells, (the cells representing the contents of four confluent Roux bottles), by mechanically shaking the cells with the antiserum for 90 minutes at 37°C followed by 18 hours (overnight) at 4°C. All the cells are removed by centrifugation at 10 000 rev/min for thirty minutes at +4°C. The absorbed serum is then tested against uninfected HeLa and Hep-2 cells to determine whether non-specific factors have been removed.

THE DETERMINATION OF THE WORKING OPTIMAL DILUTION OF CONJUGATE AND SERUM

The optimal dilutions of both conjugate and serum which gives strong fluorescence with minimal background staining must be determined. This is done by titrating dilutions of the serum against dilutions of the conjugate using a known antigen in a suitable tissue culture system. If the titre of both is reasonably high, then it may not always be necessary to absorb the materials as the non-specific effect is likely to be eliminated at the dilutions used.

REDUCTION OF NON-SPECIFIC STAINING FACTORS

Non-specific staining can be reduced by the following precautions:

1. Avoid bleeding patients immediately after meals so that the serum contains less fatty material.
2. Store materials at recommended temperatures. Conjugate and serum should be stored in small quantities at $-20°C$. Minimise repeated freezing and thawing.
3. All glass ware should be clean, and non-fluorescent. Rubber bungs must be non-toxic, preferably of white or silicone rubber.
4. Analar reagents should always be used. PBS tablets used (Oxoid Ltd.) should be adjusted between pH $7·2–7·6$.
5. Use reagents that are purified biologically, rather than by Sephadex columns or DEAE.
6. Preparations should be harvested in the most suitable conditions.
7. Antigens prepared in tissue culture—it is better to use the tissue as early as possible, i.e. just after eclipse phase of virus replication. If cytopathic effect is advanced then dead cellular material tends to flouresce non-specifically.

COVER SLIP CULTURES

Monolayer cell cultures may be conveniently grown on cover slips in a variety of ways:
(a) flying cover slips
(b) Petri dishes
(c) round cover slips in upright tubes.

The cover slips are first washed thoroughly by boiling in a weak solution of detergent (Pyroneg) for 20 minutes. They are then rinsed three times in distilled water and then three times in de-ionised water, and finally the cover slips are rinsed in 96% alcohol, 99% alcohol and then Analar acetone. The cover slips are removed singly and dried by standing upright round the edge of a dish lined with filter paper. When dry each cover slip is put into a 100×12 mm tube, the test tube rack covered with tin foil, and sterilised in a hot air oven.

The tubes are seeded with cells at a concentration suitable to give a continuous monolayer. A rubber bung is placed firmly on each tube and the rack of tubes incubated at $37°C$ on its side at an angle of $3°$ to the horizontal. The cells are incubated until a confluent sheet is obtained. It is important, particularly during the staining of the cell monolayer, to know to which side of the cover slip the cells adhere. For purposes of identification, cover slips can be purchased which are cut at one corner. The cover slips are then aligned so that the missing corners are all on the same side when viewed from one side of the rack.

THE PREPARATION OF COVER SLIP CULTURES

The cells usually take from 1–4 days incubation to become confluent. Virtually any cells can be grown in this way, e.g. RK_{13}, human embryonic lung, monkey kidney, HeLa, Hep-2, exceptions being: GOR, EB_3 and lymphblastic cell lines which grow in suspension, and do not adhere to the glass. For growth and maintenance media of the above cell lines, reference should be made to Chapter One.

An advantage of using the flying cover slips method for growing monolayers is that the cell sheet can be observed by placing the tube under the white light microscope. All cultures are screened before inoculation with virus and before fixation so that only cells in good condition are used for fluorescent staining.

Cell monolayers can be grown by inoculating a suspension of cells on top of the cover slips contained in a Petri dish using enough to flood the bottom of the dish. The Petri dish is then placed in an atmosphere containing 5% CO_2 and incubated at 37°C. The medium is changed by removing it with a Pasteur pipette and then by reflooding the dish.

An alternative method to obtain cover slips of tissue culture cells is to scrape cells gently from the sides of a tube or bottle in which they have been growing and make a suspension in PBS. A small quantity of this is spread over a cover slip and it is allowed to dry. When the PBS has evaporated a layer of cells is left which can be fixed to the cover slips by immersing in acetone. Infected layers can be obtained by the same process having grown and infected the cells on the sides of tubes or bottles and then scraping them off. The cells do, however, tend to round up and thus the morphology of the cells is not as easily seen as when the cells are grown as monolayers on cover slips.

GLASS SLIDE CULTURES

Glass slides may also be used for the growth of cell monolayers. A convenient method[12] of doing this is by growing the cells in glass rings on microslides. The glass rings are stuck to the slide by heating and then touching the surface of a mixture of equal parts paraffin wax and petroleum jelly and transferring the ring to the slide. As the wax sets the ring is stuck to the slide and forms a waterproof joint. Up to six rings may be fixed to a slide at a time. Tissue culture cells suspended in growth medium are added to each ring approximately 0·5 ml containing 50–70 000 cells. The slides, placed in Petri dishes are put in a humidified incubator at 37°C containing a flow of 5% CO_2 in air. Cell monolayers form in the rings and removal of growth medium, virus inoculation etc. are performed within the rings by a fine bore pipette attached to a source of gentle negative pressure. Before fixation of the cells, the rings are snapped off with a pair of forceps, the circular outline of the base in paraffin wax/petroleum jelly serves to delineate the area containing the cell culture for immuno-fluorescent staining.

THE INFECTION OF COVER SLIP CULTURES WITH VIRUS

The infection of cell monolayers growing on cover slips is performed as follows: the tubes are first examined to make sure the cell layers are in good condition before virus is added. The cover slips are then transferred to fresh sterile test tubes using aseptic technique. This is done so that the cells which have grown on the side of the glass tube are not competing with the cover slip monolayer for nutrients nor are they being infected with virus. 1·6 ml quantities of maintenance medium are added to these tubes and 0·4 ml of the virus suspension is added. The virus is left to absorb on to the cell sheet for 1–2 hours at room temperature, by placing the rack of tubes flat on a mechanical shaker. Gentle agitation ensures that virus particles reach the cell sheet as the medium moves over it. An alternative method of absorption is to remove the cover slips to small sterile Petri dishes. Usually about four cover slips are placed cell side up in a small Petri dish. About 0·4 ml of virus solution is placed on top of each cover slip and these are left for 1–2 hours at room temperature. After absorption each cover slip is put back into a fresh sterile test tube and 2 ml maintenance medium added. The cells are maintained in good condition by changing the medium every two days.

FIXATION

It is important to be able to locate the fluorescent viral antigen to a specific site in an individual cell, as this site is very often characteristic of the virus concerned. It is the method of fixation of the material which preserves the cellular morphology and keeps the antigen *in situ* and this is best observed as cell monolayers, or thin sections of organs.

For viral antigens, the slide or cover slip culture is immersed in acetone (analar) or ethyl alcohol, at room temperature, or +4°C for 10 minutes, after which fresh acetone is substituted and the culture stored in acetone at −20°C. An alternative method[13] is to cover the preparation with acetone at +4°C and then place this in a mixture of ethyl alcohol (95%) and solid CO_2 at a temperature of −70°C for 30 minutes. The acetone is then discarded and the cover slips removed from the tubes. The cover slips are air dried and stored at room temperature, where they may be kept for two years without deterioration. Each system studied should be tested to determine the condition required to produce optimal result.

SUMMARY

The essential requirements for the use of this technique to detect viral antigens are a good specific high titred antiserum, together with a good conjugate. The cells of the tissue culture must also be in good condition to minimise background staining caused by dead cells. The future role of the technique will, in the author's opinion, be to replace the neutralisation test and, therefore, speed diagnosis of viral infection in tissue culture; to determine the site of antigen in organ culture sections and, in many cases, it will be used as a rapid diagnosis of viral disease in specimens obtained directly from the patient.

REAGENTS AND SUPPLIERS

Chance Bros. Ltd., Smethwick 40, Birmingham.	Non-fluorescent slides and cover slips
Wellcome Reagents Ltd., Beckenham, Kent.	Liver powder
Shandon Scientific Co. Ltd., 65 Pound Lane, Willesden, London, N.W.10.	Reichert (Microscope)
Polaron Equipment Ltd., Delviljem House, Shakespeare Road, Finchley, London, N. 3.	Polarfluor B. (Mountant)
Oxoid Division, Oxo Ltd., London, S.E.1.	Phosphate Buffered Saline tablets

REFERENCES

1. COONS, A. H., CREECH, H. J., JONES, R. N. and BERLINER, E., *J. Immun.*, **45**, 159 (1942)
2. LIU, C., *Proc. Soc. rep. Biol., N.Y.*, **92**, 883 (1956)
3. HERS, J. F., *Am. Rev. resp. Dis.*, **88** (Suppl.) 333 (1963)
4. LLANES-RODAS, R. and LIU, C., *New Engl. J. Med.*, **275**, 516 (1966)
5. GARDNER, P. S. and MCQUILLAN, J., *Br. med. J.*, **3**, 340 (1968)
6. HAIRE, M., *Lancet*, **i**, 920 (1969)
7. MCQUILLAN, J. and GARDNER, P. S., *Br. med. J.*, **1**, 602 (1968)
8. WELLER, T. H. and COONS, A. H., *Proc. Soc. exp. Biol. Med.*, **86**, 789 (1954)
9. GOLDWASSER, R. A. and SHEPHARD, C. C., *J. Immun.*, **80**, 122 (1958)
10. NAIRN, R. C., *Fluorescent Protein Tracing*, 2nd edn, p. 119, Livingstone, Edinburgh (1962)
11. NAIRN, R. C., *Fluorescent Protein Tracing*, 2nd edn, p. 125, Livingstone, Edinburgh (1962)
12. SOMMERVILLE, R. G., *J. clin. Path.*, **20**, 212 (1967)
13. SANDER, G., *J. clin. Path.*, **22**, 737 (1969)

NINE

Culture of foetal cells from amniotic fluid

R. H. POULDING

Department of Pathology, Southmead Hospital, Bristol

INTRODUCTION

The intra-uterine diagnosis of many biochemical and cytogenetic defects of the foetus at an early stage of pregnancy has become possible in recent years due to the successful cultivation of foetal cells found in amniotic fluid obtained by transabdominal amniocentesis. A small proportion of the cells exfoliated into the fluid surrounding the foetus are capable of dividing and can be grown to form a monolayer of predominantly epithelial-like cells in 2–4 weeks depending on the initial concentration of viable cells. Primary cultures, or the first subcultures, are used for cytogenetic studies to determine the precise chromosome complement of the foetus, and cells harvested from a number of subcultures of the original growth are investigated biochemically to detect possible hereditary metabolic disorders particularly those associated with an enzyme deficiency. It is essential that the results of cytogenetic or biochemical analyses are available in time to permit a therapeutic abortion if so indicated. Although hypertonic saline abortions can be performed as late as the 24th week of gestation, ideally the pregnancy should be terminated not later than the 20th week[1]. Hence, allowing four weeks for culture and analysis of the cultured foetal cells (and possible repeat specimen) sampling of the amniotic fluid should be done not later than the 16th week of gestation. It has been shown[2] that the number of viable cells per unit volume of amniotic fluid increases between the 11th and 34th week and that amniocentesis is best undertaken between the 15th and 18th week when there is the greatest chance of successful culture of the foetal cells and still allows time for a therapeutic abortion.

Therefore, the main problem associated with the culture of foetal cells from amniotic fluid for intra-uterine diagnosis is the obtaining of sufficient cultured cells within a period of four weeks for chromosome or biochemical studies.

CYTOLOGY OF AMNIOTIC FLUID AT THE 16TH WEEK OF GESTATION

At the 16th week of gestation the human foetus is approximately 160 mm in length from crown to rump and weighs about 100 g. It is suspended in 100–250 ml of amniotic fluid contained by the thin transparent amnion forming the amniotic sac which provides an aquatic environment for the uniform growth of the foetus and protection against mechanical injury. The fluid is foetal in origin derived in part by the transferance of foetal extra-cellular fluid through the primitive skin and also, after the 12th week, from foetal urine. At this stage of foetal life, amniotic fluid is probably an extension of the extracellular fluid space of the foetus, subject to the same homeostatic regulations as the foetus[3].

On the foetal surface the amnion is lined with extra-embryonic ectoderm continuous with the epithelium of the skin, mucous membrane of the mouth and nasal cavities, and terminal portions of the alimentary and genito-urinary tracts of the foetus. In the early embryo, the amnio-ectodermal junction is in the ventral wall but as the foetus develops it is limited to an external covering on part of the umbilical cord. Until late in pregnancy, the amniotic ectoderm is composed of low cuboidal cells but changes to tall columnar cells after the 31st week[4]. Opaque areas in the normally transparent amnion was found to consist of layers of epithelial cells resembling areas of squamous metaplasia[5]. At the 16th week of gestation, therefore, the cell population consists largely of desquamated cells of ectodermal origin from the amniotic epithelium, the epidermis of the skin including the epithelium of the associated sweat and sebaceous glands, the epithelium of the oral cavity and parts of the alimentary, urinary and genital tracts. Also, as the foetus from an early stage swallows amniotic fluid, epithelial cells of endodermal origin from the larynx, oeso-phagus, stomach, small intestine and trachea could theoretically desquamate and form a small part of the cell population sampled by amniocentesis. The number of cells found in amniotic fluid samples show wide variation not only at different stages of gestation but even between several samples taken from the patient at the same time. Clearly much depends on the distribution of the exfoliated cells at the time of sampling. Votta[6] found a range of 400–2800 cells per ml at 28–37 weeks of gestation with a mean of 8339, and at term (37–42 weeks) a mean of 2721 cells per ml. Using the trypan blue method[7] to differentiate the viable cells in cell suspensions, Wahlström[2] found in five different cases that the number of living cells in the amniotic fluid at 16 weeks was 1940, 3720, 4500 and 5600 (mean 3955) per ml respectively. This implies that in a 10 ml sample taken at this time there will be in the order of 20 000–50 000 living cells, but as growth in amniotic fluid cell culture starts from a few scattered foci only a small proportion appear to have a growth potential.

CELL TYPES FOUND IN AMNIOTIC FLUID AT THE 16TH WEEK OF GESTATION

AMNION CELLS

Flattened, oval cuboidal up to 25 μm in diameter with a large vesicular nucleus and dense, eosinophilic cytoplasm which is often vacuolated, amnion cells occur occasionally in cell 'balls' or clusters in a fluid deposit and these are probably formed by the rolling up of a single layer of cells from the amnion lining.

SQUAMOUS CELLS FROM FOETAL EPIDERMIS

Large superficial cells, 25–80 µm in diameter with abundant eosinophilic or cyanophilic cytoplasm. Nucleus is round and frequently pyknotic. Anucleate squamous cells and cells with ghost nuclei are numerous and form the bulk of the squamous cells exfoliated into amniotic fluid.

UNDIFFERENTIATED CELL CLUSTERS

Most samples of amniotic fluid contain scanty groups of small undifferentiated cells of indeterminate origin. These groups range in size from 25 µm to 30 µm across consisting of tightly packed cells less than 10 µm in diameter, with scanty cytoplasm and round, pale staining nuclei. The most likely source of these cell clusters is the endodermal epithelium of the respiratory, alimentary or genito-urinary tracts and they are probably the cells with growth potential from which the epithelial-like growth is initiated in primary culture.

PHAGOCYTIC CELLS

Small phagocytic cells, resembling macrophages or histiocytes are found in small numbers in most samples. They can be shown to be viable and actively phagocytic whilst in stained smears and they often contain a large vacuole in the cytoplasm. These cells can either be maternal in origin or derived from the foetal mesenchyme.

ERYTHROCYTES AND LEUCOCYTES

Some samples contain a few erythrocytes and an occasional leucocyte due to contamination with maternal blood during amniocentesis.

CULTURE METHODS

The determination of the chromosomal sex of the unborn foetus by examining cells from amniotic fluid for sex chromatin was reported in 1956[8]. However, it was not until 1963 that Fuchs and Philip[9] were able to grow amniotic fluid cells but gave no details of the culture technique. Subsequently a number of different methods by various workers were published[10-16]; with the more recent ones showing a marked reduction in the time required to produce sufficient cells for chromosome studies and also, in the success rate. Basically the composite culture medium of these methods is similar—a basal defined medium such as Eagles', medium 199 or Ham's F10 supplemented with 15–30% foetal bovine serum and requiring an atmosphere of 5% CO_2 in air. The one exception is that described by Santesson[13] who used 20% pooled human serum and 5% bovine embryo extract as a supplement in place of foetal bovine serum. Bovine embryo extract and foetal bovine serum appear to contain the essential growth factor(s) for the growth of epithelial-like cell cultures providing there are sufficient living cells capable of dividing in the amniotic fluid sample. The critical cell density for initial foetal cell growth is unknown but the most rapid growth is obtained when the available cell population of the amniotic fluid sample is concentrated in one or two culture vessels with a small culture area such as a standard Leighton tube or Carrel flask. Like most cells in culture the time in which an adequate monolayer is obtained is largely dependant on the density of the cells per unit area of the culture surface of the flask. This critical cell density provides the optimum concentration of a diffusable conditioning growth factor which is directly related to the linear distance between the viable cells with growth potential[17]. To improve the plating efficiency of samples with low cell counts, Steele and Breg[10] used irradiated feeder fibroblast cell layers in

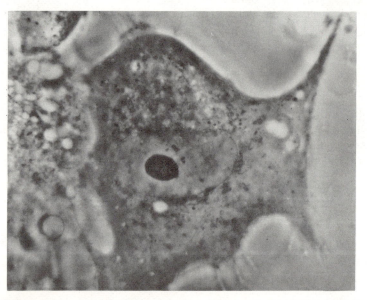

Fig. 9.1. Epithelial-like cell at the 20th day of culture. Monolayer culture initiated from amniotic fluid sample taken at the 15th week of pregnancy. (phase contrast, × 500)

the culture flasks employed for the initial inoculum of amniotic fluid cells. Fogh and Lund[18] first established the human amnion epithelial strain FL from amniotic membrane in a medium containing Earles' BSS, 20% ox serum and lactalbumin hydrolysate. Although there appears to be no definite evidence that amnion cells have a growth potential after exfoliating into amniotic fluid, the morphology of the clear, flat, epithelial-like cells of the FL amnion strain is not unlike that of the cultured cells from amniotic fluid grown in TC 199 and 30% foetal bovine serum (see Fig. 9.1). It is most likely that amniotic fluid cell cultures originate from several cell types notably amnion and undifferentiated cell clusters of endodermal origin and can give rise to both epithelial-like and fibroblast-like cell colonies. Certainly after several subcultures fibroblast-like cells predominate in rapidly growing cultures although the cells first growing in primary cultures are epithelial-like in most instances.

Of the many variants of the basic concept of culturing amniotic fluid cells those of Santesson[13] and Nadler and Gerbie[16] are selected for inclusion in this chapter because of the short time claimed by the authors to produce preparations for chromosome analysis and the high success rate of almost 100%. In addition a simple method used in this laboratory is also described which is perhaps not so rapid as the above authors' methods but has given a consistent success rate for the production of monolayers for both cytogenetic and biochemical investigations. All methods incubate cultures at 37°C in an atmosphere of 5% CO_2 in air.

METHOD A—SANTESSON *et al.* (1969)[13]

Sample size

10 ml.

Culture vessels

Carrel flasks, 50 mm in diameter, and plastic Petri dishes, 50 mm in diameter.

Culture medium

Equal parts of Hanks' BSS and Parker TC 199 with 20% of pooled human serum, 5% of bovine embryo extract (Difco) and antibiotics.

METHOD

Spin 3 ml subsamples of fluid and rinse cell deposit in the medium. Use total cell suspension from one subsample to inoculate each culture vessel. Growth is obtained after a few days and sufficient cells are available for chromosome studies or subculturing after 4–12 days incubation.

METHOD B—NADLER AND GERBIE (1970)[16]

Sample size

10 ml collected in plastic or siliconed glass tube.

Culture vessels

Small Falcon Petri dishes.

Culture medium

F-10 nutrient medium supplemented with 15% foetal bovine serum and antibiotic-antimycotic mixtures.

METHOD

After centifugation the cell pellet is resuspended in 0·5 ml foetal bovine serum and divided between five Falcon Petri dishes. The cells are immobilised by placing a coverslip over them before adding 2–3 ml of medium. This is changed every other day until colonies of cells are seen under the coverslips usually in 7–18 days. The coverslip is removed and inverted in a new Petri dish to start a new culture. Fresh medium is added to both cultures. After 24 h the coverslip is treated to obtain direct chromosome preparations and the subculture is continued to provide further monolayers for biochemical investigations.

METHOD C—GREGSON (1970)[14] WITH MODIFICATIONS

Sample size

10 ml in narrow-necked glass Universal containers.

Culture vessels

Carrel flask, 50 mm in diameter, and standard glass Leighton tube with culture area of 40×10 mm.

Culture medium

TC 199 (Wellcome)* supplemented with 30% of foetal bovine serum (Flow Laboratories).†

METHOD

1. Spin sample at 1000 rev/min for 5–10 min.

* Wellcome Reagents Limited, Beckenham, Kent.
† Flow Laboratories Limited, Irvine, Ayrshire.

2. If the cell pellet contains blood, resuspend in 1 ml of TC 199 and add 8 ml of 0·83% aqueous ammonium chloride (cooled to 4°C) to partially lyse the erythrocytes[19]. Leave at 4°C for 3 min and add 1 ml of foetal bovine serum to stop lysis of cells. Repeat centrifugation to obtain cell pellet.

3. Resuspend cell in 3 ml of medium.

4. Place 1 ml of cell suspension in a Leighton tube and the remaining 2 ml in a Carrel flask. Gas with 5% CO_2 in air and leave undisturbed for 6 days at 37°C.

5. On the 7th day, carefully decant medium into a sterile centrifuge tube from both culture flasks and replace medium. The supernatant is spun at 100 rev/min for 10 min and the deposit of cells which failed to settle is used to set up a reserve culture in a second Leighton tube.

6. In 14–21 days when the epithelial-like cells extend over a total area of approximately 200 mm^2 one of the cultures is treated with colchicine to harvest dividing cells at the metaphase. The remaining cultures are allowed to grow until sufficient cells are available to set up subcultures for biochemical and additional cytogenetic investigations.

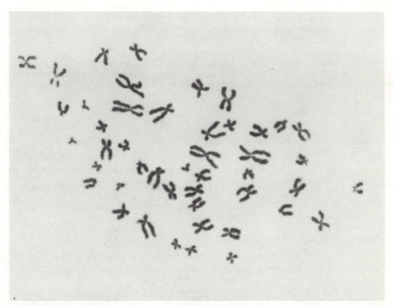

Fig. 9.2. Metaphase obtained from colchicine-treated amniotic fluid cell culture terminated on the 21st day of incubation (May-Grünwald–Giemsa stain, × 500)

The treatment of the primary cultures for chromosome preparations is the same as that described for fibroblast subcultures (see Chapter 7). A stronger trypsin solution (0·25%) is normally required to remove the cells from the growth surface and occasionally a glass scraper is necessary to assist cell detachment. The resulting cell suspension is processed through the hypotonic and fixation stages prior to staining and analysis exactly as for fibroblast cell suspensions. The number of cells from a small primary amniotic fluid cell culture is far less than that obtained from a normal fibroblast monolayer at first or second subculture. Therefore great care must be taken in the centrifugation

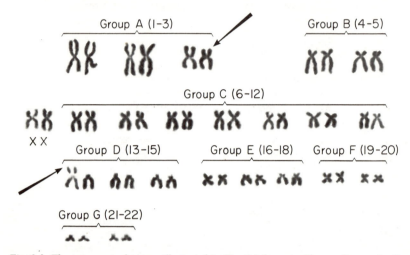

Fig. 9.3. The same metaphase as illustrated in Fig. 9.2 karyotyped according to the Denver system of classification. Note arrowed deletion of a Group A3 chromosome translocated on to the short arms of a Group D chromosome. This karyotype indicated that the foetus was a female and carried a balanced chromosomal abnormality.

and pipetting of these low cell concentrations to avoid cell loss. The quality of the chromosome preparations made by this method is comparable to those obtained from lymphocyte and fibroblast cultures—an essential factor in the detection or chromosomal structural abnormalities (Figs. 9.2 and 9.3).

REFERENCES

1. CREASEMAN, W. T., LAWRENCE, R. A. and THIEDE, H. A., *J. Am. med. Ass.*, **204**, 949 (1968)
2. WAHLSTRÖM, J., BROSSET, A. and BARTSCH, F., *Lancet*, **ii**, 1037 (1970)
3. LIND, T. and HYTTEN, F. E., *Lancet*, **i**, 1147 (1970)
4. WACHTEL, E., GORDON, H. and OLSEN, E., *J. Obstet. Gynaec. Br. Commonw.*, **76**, 596 (1969)
5. POTTER, E. L., *Pathology of the Fetus and Infant*, 2nd edn, Year Book Medical Publishers Inc., Chicago (1961)
6. VOTTA, R. A., BOBROW DE GAGNETEN, C., PARADA, O. and GIULIETTI, M., *Am. J. Obstet. Gynec.*, **102**, 571 (1968)
7. PERTOFT, H., BÄCK, O. and LINDAHL-KIESSLING, K., *Expl Cell Res.*, **50**, 355 (1968)
8. FUCHS, F. and RIS, P., *Nature, Lond.*, **177**, 330 (1956)
9. FUCHS, F. and PHILIP, J., *Nord. Med.*, **69**, 572 (1963)
10. STEELE, M. W. and BREG, W. R., *Lancet*, **i**, 383 (1966)
11. JACOBSON, C. B. and BARTER, R. H., *Am. J. Obstet. Gynec.*, **99**, 796 (1967)
12. NADLER, H. L., *Pediatrics*, **42**, 912 (1968)
13. SANTESSON, B., AKESSON, H. O., BÖÖK, J. A. and BROSSET, A., *Lancet*, **ii**, 1067 (1969)

14. GREGSON, N. M., *Lancet,* **i**, 84 (1970)
15. NELSON, M. M. and EMERY, A. E. H., *Br. med. J.,* **1**, 523 (1970)
16. NADLER, H. L. and GERBIE, A. B., *New Engl. J. Med.,* **282**, 596 (1970)
17. RUBIN, H. and REIN, A., *Growth Regulating Substances for Animal Cells in Culture*, p. 51, Wistar Institute Press, Philadelphia (1967)
18. FOGH, J. and LUND, R. O., *Proc. Soc. exp. Biol. Med.,* **94**, 532 (1957)
19. LEE, C. L. Y., GREGSON, N. M. and WALKER, S., *Lancet,* **ii**, 316 (1970)

Index